Birkhäuser

Advances in Mechanics and Mathematics

Advances in Continuum Mechanics

Volume 51

This series includes collective thematic books and monographs on advanced research topics in continuum mechanics. It is a unique forum for reference reviews discussing critically foundational issues in classical field theories under the light of up-to-date problems raised by the offspring of condensed matter physics and industrial needs, including those concerning the mechanics of microstructures and the relevant multi-field (or phase-field) representations. The series aims also to show new mathematical methods in continuum mechanics and/or recent developments of existing powerful practice. Also, the series pays attention to applications with concrete conceptual content. The series editors select editors of books on specific themes, which may split into sequential volumes.

All manuscripts are peer-reviewed to meet the highest standards of scientific literature:

- Series editors provide their final decision for inclusion in the series after manuscripts have been fully peer-reviewed;
- For contributed volumes, volume editors conduct the anonymous peer-review process before final submission;
- The peer-review process is anonymous and confidential in order to ensure reviewers provide impartial opinions

Pablo Amster • Pierluigi Benevieri
Editors

Topological Methods for Delay and Ordinary Differential Equations

With Applications to Continuum Mechanics

Birkhäuser

Editors
Pablo Amster
Departamento de Matemática
Universidad de Buenos Aires &
Imas-CONICET
Buenos Aires, Buenos Aires, Argentina

Pierluigi Benevieri
Instituto de Matemática e Estatística
Universidade de São Paulo
São Paulo, São Paulo, Brazil

ISSN 1571-8689 ISSN 1876-9896 (electronic)
Advances in Mechanics and Mathematics
ISSN 2524-4639 ISSN 2524-4647 (electronic)
Advances in Continuum Mechanics
ISBN 978-3-031-61336-4 ISBN 978-3-031-61337-1 (eBook)
https://doi.org/10.1007/978-3-031-61337-1

Mathematics Subject Classification: 34-xx, 82-xx

This book is published under the imprint Birkhäuser, www.birkhauser-science.com by the registered company Springer Nature Switzerland AG
The registered company address is: Gewerbestrasse 11, 6330 Cham, Switzerland

Preface

Continuum Mechanics is a set of space-time field theories, each a class of mathematical models that are representations of natural phenomena involving bodies extended in space and considered at a scale in which the quantum structure of the matter can be reasonably left momentarily apart. The theory of isothermal elasticity of simple bodies is a paradigmatic example; the adjective "simple" is used here in W. Noll's sense, meaning that the energy depends only on the deformation gradient. The story does not end here, of course, because Continuum Mechanics has a very broad horizon that includes even the general relativity theory. With its polymorphic nature, Continuum Mechanics is in itself a playground for researchers in mathematical analysis, differential geometry, and numerical analysis, not only in terms of applications of abstract results but—and perhaps above all—for the challenging problems frequently offered by the setting. When we look at evolutionary processes, be them described by hyperbolic or parabolic partial differential equations, and think of them in terms of numerical methods, after discretization in space, we have in the hands a system of (typically nonlinear) ordinary differential equations, an analytical structure pertaining also to lattice-type systems. Such equations (typically nonlinear and at times integro-differential in time, namely when memory effects are involved) may be analyzed at least qualitatively per se, besides proceeding with a discretization in time and the pertinent integration algorithms. When we opt for this aspect, topological methods enter into play as essential tools.

Topology has been a key tool in the study of dynamical systems since its birth in the nineteenth century with the celebrated Henri Poincare's works on celestial mechanics. Also, topological methods are essential in classifying and discussing defect structures at microscopic scales in solids or qualitatively describing peculiar features of flows pertaining to simple or complex fluids. Pertinent techniques include fixed point theory, the analysis of periodic orbits and/or chaotic attractors, and so on. Here, we focus attention on evolutionary phenomena and look at nonlinear ordinary and delay differential equations, with emphasis in the applications to mechanics and biological sciences. The book is designed for a wide audience, however interested in differential equations, functional analysis, topology and related topics, as well as in the applied sciences.

Specifically, Chap. 1 offers a clear example of how topological methods help in the analysis of ordinary differential equations. At first the focus is mainly on the history of ideas, with special attention to the celebrated Poincaré-Birkhoff theorem, from its precursors to recent developments. Then, the chapter deals with some results concerning the multiplicity of periodic solutions to a special class of Hamiltonian systems with symmetries.

Chapter 2 discusses Kepler's problem in the plane in two different relativistic versions. First, the relativistic differential operator is involved. Then, the same operator appears as a correction to the usual gravitational potential as suggested by Tullio Levi-Civita. Bifurcation results of periodic solutions for the two relativistic problems are then discussed. The approach rests on an abstract bifurcation theory developed by Alan Weinstein, applied in the case of nearly integrable Hamiltonian systems that satisfy the usual KAM isoenergetic non-degeneracy condition.

Chapter 3 collects recent results on the existence of T-periodic solutions for the relativistic pendulum for both conservative and damped circumstances. The non-variational structure of the second case implies in the application of topological degree methods a crucial difference between classical and relativistic frameworks.

The topic of T-periodic solutions reappears also in Chap. 4, dedicated to existence, multiplicity, and stability problems for a large class of micro-electro-mechanical parallel plate actuators, whose motion is governed by a Duffing equation. Mathematical tools include the method of upper and lower solutions, averaging theory and a method proposed by Rafael Ortega.

Chapter 5 deals with the analysis of chemostats linked in series. Assuming that the input of limiting nutrients is regulated by a T-periodic function, at least one T-periodic solution is obtained by means of the Poincaré translation operator. In particular, it is proven that the chain of chemostats guarantees the periodical coexistence of the microbial species.

Chapter 6 discusses a parameter-dependent thermostat problem subjected to deviated arguments. Temperature steady-states in a heated bar of finite length satisfy a second-order nonlinear differential equation. Solvability of the system rests on a recent Birkhoff-Kellogg type theorem in affine cones.

Chapter 7 deals with the analysis of forced oscillations described by non-linear implicit ordinary differential equations involving a generalized ϕ-Laplacian type differential operator. These analytical structures appear in non-Newtonian fluid-dynamics, nonlinear elasticity, diffusion of flows in porous media, theory of capillarity, and glaciology. The chapter illustrates bifurcation properties of periodic solutions, following a topological approach based on Brouwer's degree.

Chapter 8 deals also with bifurcation, now concerning a class of neutral delay differential equations, depending on a real parameter. The approach is based on a concept of oriented degree for noncompact perturbations of Fredholm maps with index zero between Banach spaces. The general result is applied to analyze Mackey–Glass-type and Duffing-type equations.

Finally, Chap. 9 shows how to combine topological analysis in phase space with dynamical systems theory. For deterministic systems, the topological structure of a flow is an invariant that provides information on the mechanisms acting in phase space to shape a flow. Appropriate topological analysis involves finding a topological representation of the underlying structure and obtaining an algebraic description, which may allow the computation of topological invariants. Developing methods to accomplish these two steps has involved several false starts as well as partially successful roads. Eventually, the attempts lead to what it is nowadays called a *templex*. The chapter introduces readers to these concepts, with emphasis on flows in phase space, the different manners of describing their topological structure, and the recently developed tools. Examples using ordinary and delay differential equations are presented.

We gratefully acknowledge Professor Paolo Maria Mariano of the University of Florence, Italy, for the invitation to edit this volume and the support he gave us throughout our work. We also thank the staff of Springer-Birkauser for the help we received and the solution to all the technical problems we encountered. In particular, we would like to mention Saveetha Balasundaram, Jeffrey Taub, Britta Rao, and Chris Eder.

Buenos Aires, Argentina
São Paulo, Brazil
November 2023

Pablo Amster
Pierluigi Benevieri

Contents

Contributors

John A. Arredondo Departamento de Matemáticas Fundación Universitaria Konrand Lorenz, Facultad de Matemáticas e Ingenierías, Bogotá, Colombia

Pierluigi Benevieri Instituto de Matemática e Estatística, Universidade de São Paulo, São Paulo, São Paulo, Brasil

Alberto Boscaggin Department of Mathematics "Giuseppe Peano", University of Torino, Torino, Italy

Alessandro Calamai Dipartimento di Ingegneria Civile, Edile e Architettura, Università Politecnica delle Marche Via Brecce Bianche, Ancona, Italy

Gisela D. Charó Centro de Investigaciones del Mar y de la Atmósfera (CIMA-FCEN, UBA), Buenos Aires, Argentina

J. Ángel Cid CITMAga, Santiago de Compostela, Spain
Departamento de Matemáticas, Universidade de Vigo, Campus de Ourense, Spain

Guillermo Contreras Colegio Pedro de Valdivia, Quilín 5473, Peñalolén–Santiago, Chile

Walter Dambrosio Department of Mathematics "Giuseppe Peano", University of Torino, Torino, Italy

Guglielmo Feltrin Department of Mathematics, Computer Science and Physics, University of Udine, Udine, Italy

Alessandro Fonda Dipartimento di Matematica e Geoscienze, Università di Trieste, Trieste, Italy

Gennaro Infante Dipartimento di Matematica e Informatica, Università della Calabria, Cosenza, Italy

Maria Patrizia Pera Dipartimento di Matematica e Informatica "Ulisse Dini", Università degli Studi di Firenze, Florence, Italy

Andrés Rivera Departamento de Ciencias Naturales y Matemáticas, Pontificia Universidad Javeriana Cali, Facultad de Ingeniería y Ciencias, Cali, Colombia

Gonzalo Robledo Departamento de Matemáticas, Universidad de Chile, Ñuñoa–Santiago, Chile

Denisse Sciamarella Institut Franco-Argentin d'Études sur le Climat et ses Impacts (IFAECI - CNRS, CONICET, IRD, UBA), Buenos Aires, Argentina

Daniel Sepúlveda Departamento de Matemática, Universidad Tecnológica Metropolitana, Ñuñoa–Santiago, Chile

Marco Spadini Dipartimento di Matematica e Informatica "Ulisse Dini", Università degli Studi di Firenze, Florence, Italy

Chapter 1
Periodic Solutions of Hamiltonian Systems with Symmetries

Alessandro Fonda

Mathematics Subject Classification 34C25

1.1 A Historical Introduction

In this section we will provide a brief historical account on the development of the theory of periodic solutions of Hamiltonian systems that originated from the celebrated Poincaré–Birkhoff Theorem. We will start from the original conjecture by Poincaré, concerning fixed points of a planar area-preserving homeomorphism, and follow the efforts done to prove it and generalize it over more than one hundred years so to obtain a powerful tool to be used in the applications. We will try to explain the relation with the Lusternik–Schnirelmann theory and its generalizations, which have been used to extend the theorem to higher dimensions, so to provide multiple periodic solutions for Hamiltonian systems. Finally, we will emphasize a recent result of the author with R. Ortega on a two-point boundary value problem, which however is strongly related with the Poincaré–Birkhoff Theorem, and which we will then use to prove the multiplicity of periodic solutions to some Hamiltonian systems with symmetries.

1.1.1 The Poincaré–Birkhoff Theorem

Among the most influential mathematicians in history, we acknowledge Jules Henri Poincaré, who passed away 112 years ago, on July 17, 1912. Just three months before his death, in the May 1912 issue of the *Rendiconti del Circolo Matematico di Palermo*, he published his paper "Sur un théorème de géométrie," which asserts

A. Fonda (✉)
Dipartimento di Matematica e Geoscienze, Università di Trieste, Trieste, Italy
e-mail: a.fonda@units.it

P. Amster, P. Benevieri (eds.), *Topological Methods for Delay and Ordinary Differential Equations*, Advances in Mechanics and Mathematics 51,
https://doi.org/10.1007/978-3-031-61337-1_1

the existence of at least two fixed points for an area-preserving homeomorphism of a planar circular annulus onto itself, such that the points of the inner circle Γ_1 are moved along Γ_1 in the clockwise sense and the points of the outer circle Γ_2 are moved along Γ_2 in the counter-clockwise sense [103].

However, Poincaré was not able to prove this result, and he tried to justify himself with these words (in our translation):

> *I have never presented such an incomplete work to the public; therefore, I think it necessary to briefly explain the reasons which convinced me to publish it, and, above all, those which drove me to start it. I have already proved in the past the existence of periodic solutions for the three body problem; however, the result was still unsatisfactory [...] While thinking at this problem, I convinced myself that the answer should depend on the truth or falseness of a geometric theorem [...] So, I was led to research the veracity of this theorem, but I met some unexpected difficulties [...] It seems that, in such a situation, I should refrain from any publication until I have solved the problem; but, after all the pointless efforts made over many months, I thought that the wiser choice was to leave the problem to mature, while resting for some years; this would have been fine if I had been sure to be able to take it up again one day; but at my age I cannot be so sure. On the other hand, the importance of the subject is too great and the quantity of results so far obtained too considerable, to resign myself to let them definitively unfruitful [...] I think that these considerations are sufficient to justify me.*

The existence of one fixed point was proved by George David Birkhoff [10] the year later, in 1913, while the proof of the existence of a second fixed point was provided by Birkhoff himself [12] only in 1925 (see also [14, 26] for a modern exposition). Since then, the "théorème de géométrie" is known as the Poincaré–Birkhoff Theorem.

Applications of the Poincaré–Birkhoff Theorem to dynamical systems coming from nonlinear mechanics and geometry were already suggested by Poincaré in [103] and studied by Birkhoff in [11, 13]. As mentioned by Zehnder in [127], Arnold considered this theorem as "the seed of symplectic topology."

1.1.2 First Extensions in the Planar Case

In the case of planar Hamiltonian systems one often looks for the existence of periodic solutions as fixed points of the Poincaré map. However, a major difficulty in the application of the Poincaré–Birkhoff Theorem is the construction of invariant annular regions. Hence, a modification of the theorem not assuming the invariance conditions for the annulus and its inner and outer boundaries became necessary for the applications.

In 1976, Jacobowitz [82, 83], following a suggestion of Moser [99], proposed a modified version of the Poincaré–Birkhoff Theorem for a topological pointed disc, showing how to apply it to the search of periodic solutions to some superlinear second-order differential equations. Applications in this direction were also given by Hartman [79] and Butler [27]. Based on Jacobowitz's result, W.-Y. Ding [41, 42] obtained a new version of the Poincaré–Birkhoff Theorem where the boundary invariance assumption was removed. Moreover, in [42] the circular annulus was replaced by a topological annulus whose inner and outer boundaries Γ_1 and Γ_2 are Jordan curves, assuming only Γ_1 to be star-shaped. It has been shown by Martins and Ureña in [94] that this star-shapedness assumption is not eliminable. Later on, Le Calvez and Wang in [88] provided an example showing that a star-shapedness assumption on the outer boundary is also needed, thus proving that Ding's theorem, hence most probably also Jacobowitz's, was not correct.

Meanwhile, Ding's theorem had been used by many authors to prove existence and multiplicity of periodic solutions of some nonautonomous Hamiltonian systems. See, e.g., [15, 16, 20, 22, 27, 28, 36–40, 46, 48, 51, 52, 56, 62, 74, 80, 82, 93, 104, 105, 109–112, 124–126]. Fortunately, all these papers deal with annuli for which both boundaries are star-shaped, and Rebelo [109] was able to save the situation proving a safer version of the Poincaré–Birkhoff Theorem where such an assumption was made. We refer to [61, 87] for further references and a complete historical account until the year 2012, the centennial of Poincaré's original paper.

Let us now state a modern version of the Poincaré–Birkhoff Theorem providing the existence and multiplicity of periodic solutions to a planar Hamiltonian system

$$x' = \partial_y H(t, x, y) , \qquad y' = -\partial_x H(t, x, y) , \tag{1.1}$$

where $H : \mathbb{R} \times \mathbb{R}^2 \to \mathbb{R}$ is a continuous function, with continuous partial derivatives $\partial_x H(t, x, y)$ and $\partial_y H(t, x, y)$. The following theorem is contained in [72].

Theorem 1.1 *Assume $H(t, x, y)$ to be T-periodic in t for some $T > 0$, and τ-periodic with respect to x, for some $\tau > 0$. Let $\gamma_1, \gamma_2 : \mathbb{R} \to \mathbb{R}$ be two continuous and T-periodic functions, with $\gamma_1(x) < \gamma_2(x)$ for every $x \in \mathbb{R}$, satisfying the following property: All solutions (x, y) of (1.1) starting with $y(0) \in [\gamma_1(x(0)), \gamma_2(x(0))]$ are defined on $[0, T]$ and are such that*

$$\begin{cases} y(0) = \gamma_1(x(0)) & \Rightarrow \quad x(T) - x(0) < 0 , \\ y(0) = \gamma_2(x(0)) & \Rightarrow \quad x(T) - x(0) > 0 . \end{cases} \tag{1.2}$$

Then, system (1.1) has at least two geometrically distinct T-periodic solutions (x, y), with $y(0) \in \,]\gamma_1(x(0)), \gamma_2(x(0))[$.

Let us specify what we mean by *geometrically distinct* solutions. By the periodicity assumption, once a solution $z(t) = (x(t), y(t))$ of (1.1) has been found, infinitely many others appear by just adding an integer multiple of τ to $x(t)$. We will call *geometrically distinct* two solutions that cannot be obtained from each other in this way.

Assumption (1.2) is usually called *twist condition*, and the same conclusion also holds if the inequalities in (1.2) are reversed. The original Poincaré's setting on an annulus can be recovered by a change of variables, choosing some suitable polar coordinates. Notice that in Theorem 1.1 no uniqueness assumption is made for the solutions of initial value problems (hence the Poincaré map could be multivalued).

As an illustrative example, consider the planar system

$$x' = \psi(y + E(t)), \qquad y' = g(x), \tag{1.3}$$

where all functions involved are continuous on the whole real line $\mathbb{R}$.

Corollary 1.1 *Let the following sign assumption hold.*

$$\exists d > 0 : \quad |\sigma| \geq d \quad \Rightarrow \quad \sigma\psi(\sigma) > 0.$$

Moreover, let the function $E(t)$ be T-periodic, and the function $g(x)$ be 2π-periodic, with $\int_0^{2\pi} g(s)\,ds = 0$. Then, system (1.3) has at least two geometrically distinct T-periodic solutions.

Proof Since $g(x)$ is bounded, Theorem 1.1 directly applies, taking as constant the functions $\gamma_1(x) = -R$ and $\gamma_2(x) = R$, with $R > 0$ large enough. □

Corollary 1.2 *Let $\phi : I \to \mathbb{R}$ be an increasing homeomorphism, with I an open interval containing 0, and $\phi(0) = 0$. Moreover, let the function $g(x)$ be continuous and 2π-periodic, with $\int_0^{2\pi} g(s)\,ds = 0$, and the function $e(t)$ be T-periodic, with $\int_0^T e(t)\,dt = 0$. Then, the equation*

$$(\phi(x'))' = g(x) + e(t) \tag{1.4}$$

has at least two geometrically distinct T-periodic solutions.

Proof Setting $\psi = \phi^{-1}$ and $E(t) = \int_0^t e(\tau)\,d\tau$, equation (1.4) can be written in the form of system (1.3), and Corollary 1.1 directly applies. □

The case $\phi(y) = y$ was first proved in [97] by a variational method, taking as a model the forced pendulum equation. For the relativistic pendulum, when $\phi(y) = y/\sqrt{1 - y^2}$, the result has been proved in [7, 25]. See also [65].

1.1.3 The Relation with Lusternik–Schnirelmann Theory

The theory of Lusternik and Schnirelmann was first published in Russian in 1930. The French translation [92] appeared in the "Exposés sur l'analyse mathématique et ses applications," published under the direction of J. Hadamard, who introduces it with the following words (our translation):

> [...] we will admire the novelty and breadth of views, the power and fecundity of ideas expressed. We considered it appropriate not to allow the reader ignore a work of this value.

As the authors say in their introduction, they were motivated by some problems raised by Poincaré in a field connecting Analysis and Topology, a domain where the most important advances at that time had been achieved by Birkhoff.

In the following years, the ideas of Lusternik and Schnirelmann were extended and generalized in several directions. In 1964, Schwartz [114] provided a first infinite-dimensional version of the theory, thus starting the development of variational methods, in view of the applications to different boundary value problems. Indeed, both ODEs and PDEs could be handled using those methods, providing several multiplicity results. In 1978, Rabinowitz [107] showed us how the periodic problem for a Hamiltonian system could be treated using a variational method. In this case, a major difficulty lies in the fact that the associated functional is strongly indefinite.

On the other hand, there have been many attempts to generalize the Poincaré–Birkhoff Theorem to higher dimensions starting with Birkhoff himself, who considered this an *outstanding question* [12, 13]. In the sixties Arnold proposed some famous conjectures, some of which are still open problems. Since then, various higher-dimensional versions of the Poincaré–Birkhoff Theorem were claimed, for maps that are close to the identity and also for monotone twist maps [2, 3, 24, 100–102, 123].

Using a different approach, Conley and Zehnder [34, Theorem 3] proved another possible version of the Poincaré–Birkhoff Theorem in higher dimensions. They obtained the multiplicity of periodic solutions for a Hamiltonian system assuming that the C^2-smooth Hamiltonian function $H = H(t, x, y)$ is periodic in t and in the space variables x_k, and quadratic in y on a neighborhood of infinity. Precisely, they assumed

$$|y| \geq R \quad \Rightarrow \quad H(t, x, y) = \tfrac{1}{2}\langle \mathbb{B}y, y\rangle + \langle a, y\rangle \,, \tag{1.5}$$

for some $R > 0$, some vector $a \in \mathbb{R}^N$, and some regular symmetric matrix $\mathbb{B}$. Remarkably, their result does not need the Poincaré time map to be close to the identity, nor to have a monotone twist.

The development of infinite-dimensional Lusternik–Schnirelmann methods would allow Szulkin [115, Theorem 4.2] to generalize the Conley and Zehnder theorem by replacing the term $\langle a, y\rangle$ by nonlinearities $G(t, x, y)$ with bounded first-order derivatives. Further results along these lines can be found in [8, 9, 31, 47, 57, 58, 75–77, 81, 84, 85, 89, 95, 96, 98, 108, 116].

1.1.4 A Higher-Dimensional Poincaré–Birkhoff Theorem for Hamiltonian Systems

Despite all this ample literature, it seems that, for the time being, there is still no genuine generalization of the Poincaré–Birkhoff theorem to higher dimensions. However, in 2017, the author together with A.J. Ureña proved a higher-dimensional version of the Poincaré–Birkhoff Theorem while considering the Hamiltonian system

$$x' = \nabla_y H(t, x, y), \qquad y' = -\nabla_x H(t, x, y), \tag{1.6}$$

where $H : \mathbb{R}\times\mathbb{R}^{2N} \to \mathbb{R}$ is a continuous function, with continuous partial gradients $\nabla_x H(t, x, y)$ and $\nabla_y H(t, x, y)$. Here,

$$x = (x_1, \dots, x_N), \qquad y = (y_1, \dots, y_N).$$

Let us see how Theorem 1.1 can be generalized in the simpler case when $\gamma_1(x)$ and $\gamma_2(x)$ are constant functions.

Theorem 1.2 *Let $H(t, x, y)$ be T-periodic in t for some $T > 0$, and that, for every $k = 1, \dots, N$, it is τ_k-periodic with respect to the variable x_k, for some $\tau_k > 0$. Let $\mathscr{D}$ be a convex body in $\mathbb{R}^N$ with a smooth boundary, and assume that all the solutions (x, y) of* (1.6) *starting with $y(0) \in \mathscr{D}$ are defined on $[0, T]$ and are such that*

$$y(0) \in \partial\mathscr{D} \quad \Rightarrow \quad \langle x(T) - x(0)\,,\, \nu_{\mathscr{D}}(y(0))\rangle > 0\,. \tag{1.7}$$

Then, system (1.6) *has at least $N + 1$ geometrically distinct T-periodic solutions (x, y), with $y(0) \in \mathrm{int}(\mathscr{D})$.*

In the above statement, $\nu_{\mathscr{D}}(y)$ denotes the outward unit normal vector to $\mathscr{D}$, for any $y \in \partial\mathscr{D}$. Theorem 1.2 extends Theorem 1.1, taking $\mathscr{D} = [\gamma_1, \gamma_2]$. The twist condition (1.7) can be further generalized in several different directions (see, e.g., [33, 53, 54, 72, 73] for the details).

This result has already found several applications (see [17, 19, 23, 29, 32, 35, 44, 45, 49, 50, 63, 64, 66–71, 78, 90, 91, 106, 118–122]). It was applied to vortex dynamics [5] and to Keplerian dynamics [18]. Finally, it has also been extended to infinite-dimensional Hamiltonian systems [21, 55].

1.1.5 A New Functional Setting

A crucial step in the application of variational methods is the choice of the space where to define the functional to be studied. For the T-periodic problem associated with the Hamiltonian system (1.6), the natural space seems to be the one made of those functions $z = (x, y)$ with both x and y belonging to $H^{1/2}_T$.

Recently the author jointly with R. Ortega [60] has obtained the following multiplicity result for a two-point boundary value problem associated with system (1.6), precisely

$$y(a) = 0 = y(b)\,. \tag{1.8}$$

Theorem 1.3 *Let $H : [a, b]\times\mathbb{R}^{2N} \to \mathbb{R}$ be a continuous function, with continuous partial gradients $\nabla_x H(t, x, y)$ and $\nabla_y H(t, x, y)$, and assume that, for every $k =$*

$1, \dots, N$, it is τ_k-periodic with respect to the variable x_k, for some $\tau_k > 0$. Assume moreover that all solutions of (1.6) *starting with $y(a) = 0$ are defined on $[a, b]$. Then, problem* (1.6)–(1.8) *has at least $N + 1$ geometrically distinct solutions.*

Surprisingly enough, *no twist condition is assumed* in the above statement. The main novelty in the proof of Theorem 1.3 lies in the fact that while for the periodic problem x and y are usually both taken in the same space $H_T^{1/2}$, in [60] we have assumed x and y to belong to some complementary spaces, which are closely related to fractional Sobolev spaces.

When the Hamiltonian function has the special form $H(t, x, y) = \frac{1}{2}|y|^2 + G(t, x)$, problem (1.6)–(1.8) becomes a Neumann boundary value problem for a second-order differential equation. We can find a multiplicity result for the Neumann problem by Castro [30] in 1980 and a similar one by Rabinowitz [108] in 1988. Both papers use variational methods.

The aim of this chapter is to provide the existence and multiplicity of T-periodic solutions for a Hamiltonian system of the type (1.6) whose Hamiltonian function is T-periodic in t and presents some symmetries. More precisely, we will assume $H(t, x, y)$ to be even, both in t and in y. In this setting, no twist condition will be needed.

We will then extend our result to infinite-dimensional systems, passing through a finite-dimensional approximation. However, the passage to the limit in the dimension will eventually only guarantee the existence of one periodic solution. We refer to [4, 21, 43, 55, 59, 86, 113] for related results in infinite dimension.

1.2 Finite-Dimensional Systems

1.2.1 The Main Result

We consider the Hamiltonian system (1.6), where $H : \mathbb{R} \times \mathbb{R}^{2N} \to \mathbb{R}$ is a continuous function, with continuous partial gradients $\nabla_x H(t, x, y)$ and $\nabla_y H(t, x, y)$. As above, we use the notation $z = (x, y)$, with

$$x = (x_1, \dots, x_N), \qquad y = (y_1, \dots, y_N).$$

Here are our assumptions:

(A_1) *The function $H(t, x, y)$ is T-periodic in the variable t, for some $T > 0$.*

(A_2) *For every $k \in \{1, \dots, N\}$, there is a $\tau_k > 0$ such that the function $H(t, x, y)$ is τ_k-periodic in the variable x_k.*

Let us now introduce the notation

$$\mathbb{T}^N = \prod_{k=1}^{N} [0, \tau_k].$$

In view of (A_2), the following assumptions will be made only for those x belonging to $\mathbb{T}^N$.

(A_3) *One has*

$$\nabla_y H\big(\tfrac{T}{2}, x, 0\big) = 0\,, \qquad \text{for every } x \in \mathbb{T}^N.$$

(A_4) *The function $H(t, x, y)$ is even in (t, y)*, i.e.,

$$H(-t, x, -y) = H(t, x, y)\,, \qquad \text{for every } (t, x, y) \in \mathbb{R} \times \mathbb{T}^N \times \mathbb{R}^N.$$

(A_5) *The solutions $z = (x, y)$ of* (1.6) *such that $z(0) \in \mathbb{T}^N \times \{0\}$ are defined on the whole time interval $\big[0, \frac{T}{2}\big]$.*

Here is our main result.

Theorem 1.4 *Let the assumptions (A_1)-(A_5) hold true. Then, system* (1.6) *has at least $N+1$ geometrically distinct T-periodic solutions (x, y). These solutions satisfy*

$$(x(-t), y(-t)) = (x(t), -y(t))\,, \qquad \textit{for every } t \in \mathbb{R}\,,$$

and

$$y\big(n\tfrac{T}{2}\big) = 0\,, \qquad \textit{for every } n \in \mathbb{Z}\,.$$

Proof By Theorem 1.3, assumptions (A_2) and (A_5) guarantee the existence of $N+1$ geometrically distinct solutions (x, y) of (1.6) satisfying the two-point boundary condition

$$y(0) = 0 = y\big(\tfrac{T}{2}\big)\,. \tag{1.9}$$

Let $z = (x, y)$ be one of such solutions, defined on the interval $\big[0, \frac{T}{2}\big]$. We will extend it to the whole line $\mathbb{R}$ so to obtain the T-periodic solution we are looking for.

Before doing this, notice that

$$x'(0) = 0 = x'\big(\tfrac{T}{2}\big)\,.$$

Indeed, by (A_4),

$$\nabla_x H(-t, x, -y) = \nabla_x H(t, x, y)\,, \qquad \nabla_y H(-t, x, -y) = -\nabla_y H(t, x, y)\,; \tag{1.10}$$

hence

$$x'(0) = \nabla_y H(0, x(0), y(0)) = \nabla_y H(0, x(0), 0) = 0\,;$$

moreover, by (A_3),

$$x'\big(\tfrac{T}{2}\big) = \nabla_y H\big(\tfrac{T}{2}, x\big(\tfrac{T}{2}\big), y\big(\tfrac{T}{2}\big)\big) = \nabla_y H\big(\tfrac{T}{2}, x\big(\tfrac{T}{2}\big), 0\big) = 0\,.$$

First of all, we extend $z(t) = (x(t), y(t))$ to the interval $\big[-\tfrac{T}{2}, \tfrac{T}{2}\big]$ by setting

$$(x(-t), y(-t)) = (x(t), -y(t))\,.$$

It is easy to see that this function is continuously differentiable on the whole interval $\big[-\tfrac{T}{2}, \tfrac{T}{2}\big]$.

Now, since $x(t)$ is even on $\big[-\tfrac{T}{2}, \tfrac{T}{2}\big]$ and $y\big(-\tfrac{T}{2}\big) = y\big(\tfrac{T}{2}\big) = 0$, we have

$$z\big(-\tfrac{T}{2}\big) = z\big(\tfrac{T}{2}\big)\,.$$

We can then extend $z(t)$ to $\mathbb{R}$ by T-periodicity, thus obtaining a continuous function. Let us prove that it is differentiable. We just need to show that it is such at $t = \frac{T}{2}$. Indeed,

$$\begin{aligned}
\lim_{t\to\frac{T}{2}^+} (x'(t), y'(t)) &= \lim_{t\to-\frac{T}{2}^+} (x'(t+T), y'(t+T)) \\
&= \lim_{t\to-\frac{T}{2}^+} (x'(t), y'(t)) \\
&= \lim_{t\to-\frac{T}{2}^+} (-x'(-t), y'(-t)) \\
&= \lim_{t\to\frac{T}{2}^-} (-x'(t), y'(t)) = \big(0, y'\big(\tfrac{T}{2}\big)\big)\,;
\end{aligned}$$

hence

$$\lim_{t\to\frac{T}{2}^+} (x'(t), y'(t)) = \lim_{t\to\frac{T}{2}^-} (x'(t), y'(t)) = \big(0, y'\big(\tfrac{T}{2}\big)\big)\,.$$

Finally, let us prove that $z(t)$ is a solution of (1.6) on the whole line $\mathbb{R}$. We know that it is a solution on $\big[0, \tfrac{T}{2}\big]$ and that it is differentiable on $\mathbb{R}$. Then, on $\big[-\tfrac{T}{2}, 0\big]$, by (1.10) we have

$$\begin{aligned}
x'(t) &= -x'(-t) = -\nabla_y H(-t, x(-t), y(-t)) \\
&= -\nabla_y H(-t, x(t), -y(t)) = \nabla_y H(t, x(t), y(t))
\end{aligned}$$

and

$$\begin{aligned}
y'(t) &= y'(-t) = -\nabla_x H(-t, x(-t), y(-t)) \\
&= -\nabla_x H(-t, x(t), -y(t)) = -\nabla_x H(t, x(t), y(t))\,.
\end{aligned}$$

We have thus proved that $z(t)$ is a solution of (1.6) on $\left[-\frac{T}{2}, 0\right]$. Hence, it is a solution on $\left[-\frac{T}{2}, \frac{T}{2}\right]$. Now, by (A_1), the T-periodicity of $z(t)$ guarantees that it is a solution on the whole $\mathbb{R}$.

We have thus proved that any solution of (1.6) on $\left[0, \frac{T}{2}\right]$ satisfying (1.9) generates a T-periodic solution of (1.6). As already said above, we have $N+1$ of them, and they are geometrically distinct. The proof is thus completed. □

We now introduce a substitute of assumption (A_5).

(A_5') There exists a constant $\bar{c} > 0$ such that

$$|\nabla_x H(t, x, y)| \leq \bar{c}(1 + |y|), \qquad \text{for every } (t, x, y) \in [0, T] \times \mathbb{T}^N \times \mathbb{R}.$$

We thus have the following.

Corollary 1.3 *Let the assumptions (A_1)-(A_4) and (A_5') hold true. Then, the same conclusion of Theorem 1.4 holds.*

Proof By the periodicity assumption in the components of the x variable, the solutions (x, y) of (1.6) can be interpreted as if $x(t)$ belongs to $\mathbb{T}^N$, for every t. By (A_5') and the compactness of $\mathbb{T}^N$, one can easily deduce that (A_5) holds, and hence Theorem 1.4 applies. □

Remark 1.1 Notice that no twist condition is needed in the above statements.

Remark 1.2 The results of this section should be compared with those in [1, 6, 117], involving Hamiltonian systems with symmetries. See also the references therein.

1.2.2 Some Corollaries

Let $N = 1$ and consider the planar system

$$x' = f(t, y), \qquad y' = g(t, x), \tag{1.11}$$

where the functions $f, g : \mathbb{R} \times \mathbb{R} \to \mathbb{R}$ are continuous. The Hamiltonian function is

$$H(t, x, y) = \int_0^y f(t, \sigma)\, d\sigma + \int_0^x g(t, s)\, ds\,.$$

Corollary 1.4 *Let the following assumptions hold true:*

$(\mathcal{A}_1)$ *f and g are T-periodic in t.*
$(\mathcal{A}_2)$ *g is τ-periodic in x, with $\int_0^\tau g(t, s)\, ds = 0$, for every t.*
$(\mathcal{A}_3)$ $f\left(\frac{T}{2}, 0\right) = 0$.
$(\mathcal{A}_4)$ *$f(-t, -y) = -f(t, y)$ and $g(-t, x) = g(t, x)$, for every t, x, y.*

Then, system (1.11) *has at least two geometrically distinct T-periodic solutions* (x, y). *These solutions satisfy*

$$(x(-t), y(-t)) = (x(t), -y(t)), \quad \text{for every } t \in \mathbb{R},$$

and

$$y\big(n\tfrac{T}{2}\big) = 0, \quad \text{for every } n \in \mathbb{Z}.$$

Proof Assumptions (A_1)–(A_4) are direct consequences of $(\mathcal{A}_1)$–$(\mathcal{A}_4)$, respectively. Moreover, assumption (A_5') is surely satisfied, since g is bounded. Hence, Corollary 1.3 applies. □

We immediately get the following result for system (1.3).

Corollary 1.5 *Assume that $E(t)$ is T-periodic and odd, with $E(\frac{T}{2}) = 0$. Let ψ be an odd function, and let $g(x)$ be 2π-periodic, with $\int_0^{2\pi} g(s)\, ds = 0$. Then, the same conclusion of Corollary 1.4 holds for system* (1.3).

Proof Setting $f(t, y) = \psi(y + E(t))$, we have

$$f(-t, -y) = \psi(-y + E(-t)) = \psi(-y - E(t)) = -\psi(y + E(t)) = -f(t, y).$$

Then, Corollary 1.4 applies. □

Let us compare Corollary 1.5 with Corollary 1.1. We notice that in Corollary 1.5 no sign assumption is needed on ψ, which however is assumed to be an odd function. Concerning the function $E(t)$, in Corollary 1.5, it is assumed to be odd, with $E(\frac{T}{2}) = 0$, while in Corollary 1.1 one only needs it to be T-periodic.

1.3 Infinite-Dimensional Systems

In this section we deal with an infinite-dimensional Hamiltonian system on a separable real Hilbert space $\mathcal{H}$. Precisely, we consider the system (1.6), where $H : \mathbb{R} \times \mathcal{H} \times \mathcal{H} \to \mathbb{R}$ is assumed to be continuous in (t, x, y) and continuously differentiable in $z = (x, y)$.

It will often be convenient to identify the space $\mathcal{H}$ with ℓ^2, the space of real sequences $\xi = (\xi_k)_{k\geq 1}$ such that $\sum_{k=1}^{\infty} \xi_k^2 < \infty$, endowed with the usual scalar product

$$\langle \xi, \tilde{\xi} \rangle_{\mathcal{H}} = \sum_{k=1}^{\infty} \xi_k \tilde{\xi}_k,$$

and the associated norm $\|\xi\|_{\mathcal{H}} = \sqrt{\langle \xi, \xi \rangle_{\mathcal{H}}}$. We will also denote by $\|\cdot\|_{\mathcal{H}\times\mathcal{H}}$ the usual norm in the product space $\mathcal{H} \times \mathcal{H}$.

We can thus rewrite (1.6) as a system of infinitely many scalar ODEs, i.e.,

$$\begin{cases} x_k' = \dfrac{\partial H}{\partial y_k}(t, (x_1, x_2, \dots), (y_1, y_2, \dots)), \\ y_k' = -\dfrac{\partial H}{\partial x_k}(t, (x_1, x_2, \dots), (y_1, y_2, \dots)), \end{cases} \qquad k = 1, 2, \dots, \tag{1.12}$$

where $x = (x_1, x_2, \dots)$ and $y = (y_1, y_2, \dots)$ belong to ℓ^2.

Here are our assumptions:

($\mathscr{A}_1$) *The function $H(t, x, y)$ is T-periodic in the variable t, for some $T > 0$.*

($\mathscr{A}_2$) *There exists a sequence of positive real numbers $(\tau_k)_{k\geq 1}$ in ℓ^2 such that, for every $k \geq 1$, the function $H(t, x, y)$ is τ_k-periodic in the variable x_k.*

It will be convenient to use the notation

$$\mathbb{T}^\infty = \prod_{k=1}^{\infty} [0, \tau_k].$$

This is a compact set homeomorphic to the Hilbert cube. In view of ($\mathscr{A}_2$), the following assumptions will be made only for those x belonging to $\mathbb{T}^\infty$.

($\mathscr{A}_3$) *One has*

$$\nabla_y H\big(\tfrac{T}{2}, x, 0\big) = 0, \qquad \text{for every } x \in \mathbb{T}^\infty.$$

($\mathscr{A}_4$) *The function $H(t, x, y)$ is even in (t, y)*, i.e.,

$$H(-t, x, -y) = H(t, x, y), \qquad \text{for every } (t, x, y) \in [0, T] \times \mathbb{T}^\infty \times \mathcal{H}.$$

($\mathscr{A}_5$) There exists a constant $\bar{c} > 0$ such that

$$|\nabla_x H(t, x, y)| \leq \bar{c}(1 + |y|), \qquad \text{for every } (t, x, y) \in [0, T] \times \mathbb{T}^\infty \times \mathcal{H}.$$

($\mathscr{A}_6$) *There exists a constant $L > 0$ such that*

$$\|\nabla_z H(t, z_1) - \nabla_z H(t, z_2)\|_{\mathcal{H}\times\mathcal{H}} \leq L\, \|z_1 - z_2\|_{\mathcal{H}\times\mathcal{H}},$$
$$\text{for every } t \in [0, T] \text{ and } z_1, z_2 \in \mathbb{T}^\infty \times \overline{B}(0, \bar{c}Te^{\bar{c}T}).$$

(We have denoted by $\overline{B}(0, R)$ the closed ball in $\mathcal{H}$ centered at 0 with radius R.)

Here is our result.

Theorem 1.5 *Let the assumptions ($\mathscr{A}_1$)-($\mathscr{A}_6$) hold true. Then, system* (1.12) *has a T-periodic solution. This solution (x, y) satisfies*

$$(x(-t), y(-t)) = (x(t), -y(t)), \qquad \textit{for every } t \in \mathbb{R},$$

and

$$y\big(n\tfrac{T}{2}\big) = 0\,, \qquad \textit{for every } n \in \mathbb{Z}\,.$$

Remark 1.3 This result should be compared with [21, Theorem 2.1], where a twist condition involving some finite-dimensional approximating systems was required. Here we do not need any twist condition. The same as in [21], the multiplicity of solutions remains an open problem.

Proof We proceed as in [21]. First of all we notice that any solution $z(t) = (x(t), y(t))$ of (1.12) starting with $z(0) \in \mathbb{T}_\infty \times \{0\}$ is defined on $[0, T]$ and satisfies

$$\|y(t)\| \leq \bar{c}Te^{\bar{c}T}\,, \qquad \text{for every } t \in [0, T]\,.$$

This follows from the compactness of $\mathbb{T}^\infty$, the linear growth assumption ($\mathscr{A}_5$), and Gronwall's Lemma. (Here, and in the sequel of the proof, we use a simplified notation for the norms.) Moreover, by ($\mathscr{A}_6$), such a solution is therein unique.

For every integer $N \geq 1$ we define the projection $P_N : \ell^2 \to \mathbb{R}^N$ as

$$P_N(\xi_1, \xi_2, \dots) = (\xi_1, \xi_2, \dots, \xi_N)$$

and the immersion $I_N : \mathbb{R}^N \to \ell^2$ as

$$I_N(\eta_1, \eta_2, \dots, \eta_N) = (\eta_1, \eta_2, \dots, \eta_N, 0, 0, \dots)\,.$$

Let $H_N : \mathbb{R} \times \mathbb{R}^N \times \mathbb{R}^N \to \mathbb{R}$ be defined as

$$H_N(t, u, v) = H(t, I_N u, I_N v)\,, \tag{1.13}$$

and consider the Hamiltonian system

$$u' = \nabla_v H_N(t, u, v)\,, \qquad v' = -\nabla_u H_N(t, u, v)\,, \tag{1.14}$$

where $u = (u_1, \dots, u_N)$ and $v = (v_1, \dots, v_N)$ belong to $\mathbb{R}^N$. Since

$$\nabla_u H_N(t, u, v) = P_N \nabla_x H(t, I_N u, I_N v)\,,$$
$$\nabla_v H_N(t, u, v) = P_N \nabla_y H(t, I_N u, I_N v)\,,$$

it can be verified that, for N large enough, all the assumptions of Corollary 1.3 hold true for system (1.14). Hence, there exists a T-periodic solution $w^N(t) = (u^N(t), v^N(t))$ of (1.14) with $w^N(0) = w^N\big(\frac{T}{2}\big) \in \mathbb{T}^N \times \{0\}$, and

$$\|I_N v^N(t)\| \leq \bar{c}Te^{\bar{c}T}\,, \qquad \text{for every } t \in [0, T]\,.$$

We now define the operators $\mathscr{I}_N : \mathbb{R}^N \times \mathbb{R}^N \to \ell^2 \times \ell^2$ as

$$\mathscr{I}_N(u, v) = (I_N u, I_N v) = ((u_1, \ldots, u_N, 0, \ldots), (v_1, \ldots, v_N, 0, \ldots)),$$

and set $z_0^N = \mathscr{I}_N w^N(0)$. Thus, we have a sequence $(z_0^N)_N$ in the compact set $\mathbb{T}^\infty \times \{0\}$, which therefore has a subsequence, still denoted by $(z_0^N)_N$, which converges to some $z_0 \in \mathbb{T}^\infty \times \{0\}$. In view of the arguments at the beginning of the proof, the solution $z(t)$ of (1.6) starting from $z(0) = z_0$ is uniquely defined on $[0, T]$. Hence, following the lines of the proof in [21, Theorem 2.1], it is possible to prove that

$$\lim_{N\to\infty} \mathscr{I}_N w^N(t) = z(t), \quad \text{uniformly in } t \in [0, T].$$

Being $w^N(0) = w^N(T)$, we have $z(0) = z(T)$. Extending $z(t)$ by T-periodicity, by $(\mathscr{A}_1)$ we have thus found the periodic solution we were looking for. □

References

1. A. Ambrosetti, V. Coti Zelati, I. Ekeland, Symmetry breaking in Hamiltonian systems. J. Differ. Equ. **67**, 165–184 (1987)
2. V.I. Arnold, *Mathematical Methods in Classical Mechanics* (Springer, Berlin, 1978) (translated from the 1974 Russian edition)
3. A. Banyaga, On fixed points of symplectic maps. Inventiones Math. **56**, 215–229 (1980)
4. V. Barbu, Abstract periodic Hamiltonian systems. Adv. Differ. Equ. **1**, 675–688 (1996)
5. T. Bartsch, M. Sacchet, Periodic solutions with prescribed minimal period of vortex type problems in domains. Nonlinearity **31**, 2156–2172 (2018)
6. T. Bartsch, Z.-Q. Wang, Periodic solutions of spatially periodic, even Hamiltonian systems. J. Differ. Equ. **135**, 103–128 (1997)
7. C. Bereanu, P.J. Torres, Existence of at least two periodic solutions of the forced relativistic pendulum. Proc. Am. Math. Soc. **140**, 2713–2719 (2012)
8. O. Bernardi, F. Cardin, Poincaré–Birkhoff periodic orbits for mechanical Hamiltonian systems on $T^*\mathbb{T}^n$. J. Math. Phys. **47**, 15 pp. (2006)
9. M.L. Bertotti, E. Zehnder, A Poincaré–Birkhoff-type result in higher dimensions. *Stochastic Processes in Classical and Quantum Systems (Ascona, 1985)*. Lecture Notes in Phys. vol. 262 (Springer, Berlin, 1986), pp. 54–64
10. G.D. Birkhoff, Proof of Poincaré's geometric theorem. Trans. Am. Math. Soc. **14**, 14–22 (1913)
11. G.D. Birkhoff, Dynamical systems with two degrees of freedom. Trans. Am. Math. Soc. **18**, 199–300 (1917)
12. G.D. Birkhoff, An extension of Poincaré's last geometric theorem. Acta Math. **47**, 297–311 (1925)
13. G.D. Birkhoff, Une généralisation à n dimensions du dernier théorème de géométrie de Poincaré. C. R. Acad. Sci. Paris **192**, 196–198 (1931)
14. G.D. Birkhoff, *Dynamical Systems* (Amer. Math. Soc., New York, 1927)
15. A. Boscaggin, Subharmonic solutions of planar Hamiltonian systems: a rotation number approach. Adv. Nonlin. Stud. **11**, 77–103 (2011)
16. A. Boscaggin, Periodic solutions to superlinear planar Hamiltonian systems. Port. Math. **69**, 127–140 (2012)

17. A. Boscaggin, W. Dambrosio, A note on the existence of multiple solutions for a class of systems of second order ODEs. J. Math. Anal. Appl. **415**, 610–622 (2014)
18. A. Boscaggin, W. Dambrosio, G. Feltrin, Periodic solutions to a perturbed relativistic Kepler problem. SIAM J. Math. Anal. **53**, 5813–5834 (2021)
19. A. Boscaggin, G. Feltrin, Positive periodic solutions to an indefinite Minkowski-curvature equation. J. Differ. Equ. **269**, 5595–5645 (2020)
20. A. Boscaggin, A. Fonda, M. Garrione, A multiplicity result for periodic solutions of second order differential equations with a singularity. Nonlinear Anal. **75**, 4457–470 (2012)
21. A. Boscaggin, A. Fonda, M. Garrione, An infinite-dimensional version of the Poincaré–Birkhoff theorem on the Hilbert cube. Ann. Sc. Norm. Super. Pisa Cl. Sci. (5) **20**, 751–770 (2020)
22. A. Boscaggin, M. Garrione, Resonance and rotation numbers for planar Hamiltonian systems: multiplicity results via the Poincaré–Birkhoff theorem. Nonlinear Anal. **74**, 4166–4185 (2011)
23. A. Boscaggin, E. Muñoz-Hernández, Planar Hamiltonian systems: index theory and applications to the existence of subharmonics. Nonlinear Anal. **226**, Paper No. 113142, 35 pp. (2023)
24. A. Boscaggin, R. Ortega, Monotone twist maps and periodic solutions of systems of Duffing type. Math. Proc. Cambridge Philos. Soc. **157**, 279–296 (2014)
25. H. Brezis, J. Mawhin, Periodic solutions of the forced relativistic pendulum. Differ. Integr. Equ. **23**, 801–810 (2010)
26. M. Brown, W.D. Neumann, Proof of the Poincaré–Birkhoff fixed point theorem. Michigan Math. J. **24**, 21–31 (1977)
27. G.J. Butler, The Poincaré–Birkhoff "twist" theorem and periodic solutions of second-order nonlinear differential equations, in *Differential Equations* (Proc. Eighth Fall Conf., Oklahoma State Univ., Stillwater, Okla., 1979) (Academic Press, New York, 1980), pp. 135–147
28. P. Buttazzoni, A. Fonda, Periodic perturbations of scalar second order differential equations. Discrete Contin. Dyn. Syst. **3**, 451–455 (1997)
29. A. Calamai, A. Sfecci, Multiplicity of periodic solutions for systems of weakly coupled parametrized second order differential equations. NoDEA Nonlinear Differ. Equ. Appl. **24**, Paper No. 4, 17 pp. (2017)
30. A. Castro, Periodic solutions of the forced pendulum equation. *Differential Equations* (Proc. Eighth Fall Conf., Oklahoma State Univ., Stillwater, 1979) (Academic Press, New York, 1980), pp. 149–160
31. K.C. Chang, On the periodic nonlinearity and the multiplicity of solutions. Nonlinear Anal. **13**, 527–537 (1989)
32. Z. Cheng, J. Ren, Existence of multiplicity harmonic and subharmonic solutions for second-order quasilinear equation via Poincaré–Birkhoff twist theorem. Math. Methods Appl. Sci. **40**, 6801–6822 (2017)
33. F. Chen, D. Qian, An extension of the Poincaré–Birkhoff theorem for Hamiltonian systems coupling resonant linear components with twisting components. J. Differ. Equ. **321**, 415–448 (2022)
34. C.C. Conley, E.J. Zehnder, The Birkhoff–Lewis fixed point theorem and a conjecture of V.I. Arnold. Invent. Math. **73**, 33–49 (1983)
35. C. Corsato, P. Omari, F. Zanolin, Subharmonic solutions of the prescribed curvature equation. Commun. Contemp. Math. **18**, 1550042, 33 pp. (2016)
36. F. Dalbono, C. Rebelo, Poincaré–Birkhoff fixed point theorem and periodic solutions of asymptotically linear planar Hamiltonian systems. *Turin Fortnight Lectures on Nonlinear Analysis*, vol. 60(2002) (2001) (Rend. Sem. Mat. Univ. Politec. Torino, 2003), pp. 233–263
37. M.A. Del Pino, R.F. Manásevich, A. Murua, On the number of 2π-periodic solutions for $u'' + g(u) = s(1 + h(t))$ using the Poincaré–Birkhoff Theorem. J. Differ. Equ. **95**, 240–258 (1992)
38. T. Ding, An infinite class of periodic solutions of periodically perturbed Duffing equations at resonance. Proc. Am. Math. Soc. **86**, 47–54 (1982)

39. T. Ding, F. Zanolin, Periodic solutions of Duffing's equations with superquadratic potential. J. Differ. Equ. **95**, 240–258 (1992)
40. T. Ding, F. Zanolin, Subharmonic solutions of second order nonlinear equations: a time-map approach. Nonlinear Anal. **20**, 509–532 (1993)
41. W.-Y. Ding, Fixed points of twist mappings and periodic solutions of ordinary differential equations. Acta Math. Sinica **25**, 227–235 (1982) (In Chinese)
42. W.-Y. Ding, A generalization of the Poincaré–Birkhoff theorem. Proc. Am. Math. Soc. **88**, 341–346 (1983)
43. G. Dincă, D. Paşca, Existence theorem of periodical solutions of Hamiltonian systems in infinite-dimensional Hilbert spaces. Differ. Integr. Equ. **14**, 405–426 (2001)
44. T. Dondè, F. Zanolin, Multiple periodic solutions for one-sided sublinear systems: a refinement of the Poincaré–Birkhoff approach. Topol. Methods Nonlinear Anal. **55**, 565–581 (2020)
45. C. Fabry, A. Fonda, A systematic approach to nonresonance conditions for periodically forced planar Hamiltonian systems. Ann. Mat. Pura Appl. (4) **201**, 1033–1074 (2022)
46. C. Fabry, J. Mawhin, Oscillations of a forced asymmetric oscillator at resonance. Nonlinearity **13**, 493–505 (2000)
47. P.L. Felmer, Periodic solutions of spatially periodic Hamiltonian systems. J. Differ. Equ. **98**, 143–168 (1992)
48. A. Fonda, Positively homogeneous Hamiltonian systems in the plane. J. Differ. Equ. **200**, 162–184 (2004)
49. A. Fonda, M. Garrione, P. Gidoni, Periodic perturbations of Hamiltonian systems. Adv. Nonlinear Anal. **5**, 367–382 (2016)
50. A. Fonda, M. Garzón, A. Sfecci, An extension of the Poincaré–Birkhoff Theorem coupling twist with lower and upper solutions. J. Math. Anal. Appl. **528**, Paper No. 127599, 33 pp. (2023)
51. A. Fonda, L. Ghirardelli, Multiple periodic solutions of scalar second order differential equations. Nonlinear Anal. **72**, 4005–4015 (2010)
52. A. Fonda, L. Ghirardelli, Multiple periodic solutions of Hamiltonian systems in the plane. Topol. Meth. Nonlin. Anal. **36**, 27–38 (2010)
53. A. Fonda, P. Gidoni, An avoiding cones condition for the Poincaré–Birkhoff Theorem. J. Differ. Equ. **262**, 1064–1084 (2017)
54. A. Fonda, P. Gidoni, Coupling linearity and twist: an extension of the Poincaré–Birkhoff theorem for Hamiltonian systems. NoDEA Nonlinear Differ. Equ. Appl. **27**, Paper No. 55, 26 pp. (2020)
55. A. Fonda, G. Klun, A. Sfecci, Periodic solutions of nearly integrable Hamiltonian systems bifurcating from infinite-dimensional tori. Nonlinear Anal. **201**, 111720, 16 pp. (2020)
56. A. Fonda, R.F. Manasevich, F. Zanolin, Subharmonic solutions for some second order differential equations with singularities. SIAM J. Math. Anal. **24**, 1294–1311 (1993)
57. A. Fonda, J. Mawhin, Multiple periodic solutions of conservative systems with periodic nonlinearity, in *Differential Equations and Applications* (Columbus, 1988) (Ohio Univ. Press, Athens, 1989), pp. 298–304
58. A. Fonda, J. Mawhin, Critical point theory and multiple periodic solutions of conservative systems with periodic nonlinearity, in *The Problem of Plateau* (World Sci. Publ., River Edge, 1992), pp. 111–128
59. A. Fonda, J. Mawhin, M. Willem, Multiple periodic solutions of infinite-dimensional pendulum-like equations. Pure Appl. Funct. Anal. **5**, 951–963 (2020)
60. A. Fonda, R. Ortega, A two-point boundary value problem associated with Hamiltonian systems on a cylinder. Rend. Circ. Mat. Palermo. Online first. https://doi.org/10.1007/s12215-023-00872-w
61. A. Fonda, M. Sabatini, F. Zanolin, Periodic solutions of perturbed Hamiltonian systems in the plane by the use of the Poincaré–Birkhoff Theorem. Topol. Meth. Nonlinear Anal. **40**, 29–52 (2012)

62. A. Fonda, Z. Schneider, F. Zanolin, Periodic oscillations for a nonlinear suspension bridge model. J. Comp. Appl. Math. **52**, 113–140 (1994)
63. A. Fonda, A. Sfecci, Periodic solutions of weakly coupled superlinear systems. J. Differ. Equ. **260**, 2150–2162 (2016)
64. A. Fonda, A. Sfecci, Multiple periodic solutions of Hamiltonian systems confined in a box. Discrete Contin. Dyn. Syst. **37**, 1425–1436 (2017)
65. A. Fonda, R. Toader, Periodic solutions of pendulum-like Hamiltonian systems in the plane. Adv. Nonlin. Stud. **12**, 395–408 (2012)
66. A. Fonda, R. Toader, Subharmonic solutions of Hamiltonian systems displaying some kind of sublinear growth. Adv. Nonlinear Anal. **8**, 583–602 (2019)
67. A. Fonda, R. Toader, Subharmonic solutions of weakly coupled Hamiltonian systems. J. Dynam. Differ. Equ. **35**, 2337–2353 (2023)
68. A. Fonda, P.J. Torres, Periodic solutions of discontinuous second order differential equations. The porpoising effect. Nonlinear Anal. Real World Appl. **74**, Paper No. 103948, 13 pp. (2023)
69. A. Fonda, W. Ullah, Periodic solutions of Hamiltonian systems coupling twist with generalized lower/upper solutions (2022). Preprint
70. A. Fonda, W. Ullah, Periodic solutions of Hamiltonian systems coupling twist with an isochronous center (2023). Preprint
71. A. Fonda, W. Ullah, Boundary value problems associated with Hamiltonian systems coupled with positively-(p, q)-homogeneous systems (2023). Preprint
72. A. Fonda, A.J. Ureña, A higher dimensional Poincaré–Birkhoff theorem for Hamiltonian flows. Ann. Inst. H. Poincaré Anal. Non Linéaire **34**, 679–698 (2017)
73. A. Fonda, A.J. Ureña, A Poincaré–Birkhoff theorem for Hamiltonian flows on nonconvex domains. J. Math. Pures Appl. (9) **129**, 131–152 (2019)
74. A. Fonda, F. Zanolin, Periodic oscillations of forced pendulums with a very small length. Proc. Roy. Soc. Edinburgh **127A**, 67–76 (1997)
75. J. Franks, Generalizations of the Poincaré–Birkhoff theorem. Ann. Math. **128**, 139–151 (1988)
76. J. Franks, Erratum to “Generalizations of the Poincaré–Birkhoff theorem”. Ann. Math. **164**, 1097–1098 (2006)
77. G. Fournier, D. Lupo, M. Ramos, M. Willem, Limit relative category and critical point theory, in: Dynamics reported. *Expositions in Dynamical Systems* (vol. 3) (Springer, Berlin, 1994), pp. 1–24
78. P. Gidoni, A. Margheri, Lower bounds on the number of periodic solutions for asymptotically linear planar Hamiltonian systems. Discrete Contin. Dyn. Syst. **39**, 585–606 (2019)
79. P. Hartman, On boundary value problems for superlinear second order differential equations. J. Differ. Equ. **26**, 37–53 (1977)
80. A.R. Hausrath, R.F. Manásevich, Periodic solutions of a periodically perturbed Lotka-Volterra equation using the Poincaré–Birkhoff theorem. J. Math. Anal. Appl. **157**, 1–9 (1991)
81. N. Hingston, Subharmonic solutions of Hamiltonian equations on tori. Ann. Math. (2) **170**, 529–560 (2009)
82. H. Jacobowitz, Periodic solutions of $x'' + f(t, x) = 0$ via the Poincaré–Birkhoff theorem. J. Differ. Equ. **20**, 37–52 (1976)
83. H. Jacobowitz, Corrigendum: The existence of the second fixed point: A correction to “Periodic solutions of $x'' + f(t, x) = 0$ via the Poincaré–Birkhoff theorem”. J. Differ. Equ. **25**, 148–149 (1977)
84. P. Jebelean, J. Mawhin, C. Şerban, A vector p-Laplacian type approach to multiple periodic solutions for the p-relativistic operator. Commun. Contemp. Math. **19**, 1650029, 16 pp. (2017)
85. F. Josellis, Lusternik–Schnirelmann theory for flows and periodic orbits for Hamiltonian systems on $\mathbb{T}^n \times \mathbb{R}^n$. Proc. Lond. Math. Soc. (3) **68**, 641–672 (1994)
86. S.B. Kuksin, Nearly integrable infinite-dimensional Hamiltonian systems. *Lecture Notes in Math.* No. 1556 (Springer, Berlin, 1993)

87. P. Le Calvez, About Poincaré–Birkhoff theorem. Publ. Mat. Urug. **13**, 61–98 (2011)
88. P. Le Calvez, J. Wang, Some remarks on the Poincaré–Birkhoff theorem. Proc. Am. Math. Soc. **138**, 703–715 (2010)
89. J.Q. Liu, A generalized saddle point theorem. J. Differ. Equ. **82**, 372–385 (1989)
90. J. López-Gómez, E. Muñoz-Hernández, F. Zanolin, On the applicability of the Poincaré–Birkhoff twist theorem to a class of planar periodic predator-prey models. Discrete Contin. Dyn. Syst. **40**, 2393–2419 (2020)
91. J. López-Gómez, E. Muñoz-Hernández, F. Zanolin, The Poincaré–Birkhoff theorem for a class of degenerate planar Hamiltonian systems. Adv. Nonlinear Stud. **21**, 489–499 (2021)
92. L. Lusternik, L. Schnirelmann, *Méthodes topologiques dans les problèmes variationnels* (Hermann, Paris, 1934)
93. A. Margheri, C. Rebelo, F. Zanolin, Maslov index, Poincaré–Birkhoff theorem and periodic solutions of asymptotically linear planar Hamiltonian systems. J. Differ. Equ. **183**, 342–367 (2002)
94. R. Martins, A.J. Ureña, The star-shaped condition on Ding's version of the Poincaré–Birkhoff theorem. Bull. Lond. Math. Soc. **39**, 803–810 (2007)
95. J. Mawhin, Forced second order conservative systems with periodic nonlinearity, Analyse non linéaire (Perpignan, 1987). Ann. Inst. H. Poincaré Anal. Non Linéaire **6**(suppl.), 415–434 (1989)
96. J. Mawhin, Multiplicity of solutions of variational systems involving ϕ-Laplacians with singular ϕ and periodic nonlinearities. Discrete Contin. Dyn. Syst. **32**, 4015–4026 (2012)
97. J. Mawhin, M. Willem, Multiple solutions of the periodic boundary value problem for some forced pendulum-type equations. J. Differ. Equ. **52**, 264–287 (1984)
98. J. Mawhin, M. Willem, Variational methods and boundary value problems for vector second order differential equations and applications to the pendulum equation. Nonlinear analysis and optimization (Bologna, 1982). *Lecture Notes in Math.* No. 1107 (Springer, Berlin, 1984), pp. 181–192
99. J. Moser, *Mimeographed Lecture Notes on Ordinary Differential Equations* (New York University, New York, 1969)
100. J. Moser, *Stable and Random Motions in Dynamical Systems* (Princeton University Press, Princeton, 1973)
101. J. Moser, Monotone twist mappings and the calculus of variations. Ergodic Theory Dynam. Syst. **6**, 401–413 (1986)
102. J. Moser, E.J. Zehnder, Notes on Dynamical Systems. *Courant Lecture Notes in Mathematics* (vol. 12) (Amer. Math. Soc., Providence, 2005)
103. H. Poincaré, Sur un théorème de géométrie. Rend. Circ. Mat. Palermo **33**, 375–407 (1912)
104. D. Qian, Infinity of subharmonics for asymmetric Duffing equations with the Lazer–Leach–Dancer condition. J. Differ. Equ. **171**, 233–250 (2001)
105. D. Qian, P.J. Torres, Periodic motions of linear impact oscillators via the successor map. SIAM J. Math. Anal. **36**, 1707–1725 (2005)
106. D. Qian, P.J. Torres, P. Wang, Periodic solutions of second order equations via rotation numbers. J. Differ. Equ. **266**, 4746–4768 (2019)
107. P.H. Rabinowitz, Periodic solutions of Hamiltonian systems. Comm. Pure Appl. Math. **31**, 157–184 (1978)
108. P.H. Rabinowitz, On a class of functionals invariant under a $\mathbb{Z}^n$ action. Trans. Am. Math. Soc. **310**, 303–311 (1988)
109. C. Rebelo, A note on the Poincaré–Birkhoff fixed point theorem and periodic solutions of planar systems. Nonlinear Anal. **29**, 291–311 (1997)
110. C. Rebelo, Multiple periodic solutions of second order equations with asymmetric nonlinearities. Discrete Cont. Dynam. Syst. **3**, 25–34 (1997)
111. C. Rebelo, F. Zanolin, Multiplicity results for periodic solutions of second order ODEs with asymmetric nonlinearities. Trans. Am. Math. Soc. **348**, 2349–2389 (1996)
112. C. Rebelo, F. Zanolin, Multiple periodic solutions for a second order equation with one-sided superlinear growth. Dynam. Contin. Discrete Impuls. Syst. **2**, 1–27 (1996)

113. R. Schmid, Infinite-dimensional Hamiltonian Systems. *Monographs and Textbooks in Physical Science* (Bibliopolis, Naples, 1987)
114. J.T. Schwartz, Generalizing the Lusternik–Schnirelman theory of critical points. Comm. Pure Appl. Math. **17**, 307–315 (1964)
115. A. Szulkin, A relative category and applications to critical point theory for strongly indefinite functionals. Nonlinear Anal. **15**, 725–739 (1990)
116. A. Szulkin, Cohomology and Morse theory for strongly indefinite functionals. Math. Z. **209**, 375–418 (1992)
117. A. Vanderbauwhede, Local Bifurcation and Symmetry. *Research Notes in Math* (Pitman, New York, 1982)
118. S. Wang, Periodic solutions of weakly coupled superlinear systems with indefinite terms. NoDEA Nonlinear Differ. Equ. Appl. **29**, Paper No. 36, 22 pp. (2022)
119. Z. Wang, T. Ma, Periodic solutions of planar Hamiltonian systems with asymmetric nonlinearities. Bound. Value Probl. **2017**, Paper No. 46, 16 pp. (2017)
120. Z. Wang, T. Ma, Infinitely many periodic solutions of planar Hamiltonian systems via the Poincaré–Birkhoff theorem. Bound. Value Probl. **2018**, Paper No. 102, 19 pp. (2018)
121. S. Wang, D. Qian, Subharmonic solutions of indefinite Hamiltonian systems via rotation numbers. Adv. Nonlinear Stud. **21**, 557–578 (2021)
122. S. Wang, D. Qian, Periodic solutions of p-Laplacian equations via rotation numbers. Commun. Pure Appl. Anal. **20**, 2117–2138 (2021)
123. A. Weinstein, Lectures on Symplectic Manifolds. *Reg. Conf. Series Math.* (vol. 29) (Amer. Math. Soc., Providence, 1977)
124. P. Yan, M. Zhang, Rotation number, periodic Fučík spectrum and multiple periodic solutions. Commun. Contemp. Math. **12**, 437–455 (2010)
125. C. Zanini, Rotation numbers, eigenvalues, and the Poincaré–Birkhoff theorem. J. Math. Anal. Appl. **279**, 290–307 (2003)
126. C. Zanini, F. Zanolin, A multiplicity result of periodic solutions for parameter dependent asymmetric non-autonomous equations. Dyn. Contin. Discrete Impuls. Syst. Ser. A Math. Anal. **12**, 343–361 (2005)
127. E.J. Zehnder, The beginnings of symplectic topology in Bochum in the early eighties. Jahresber. Dtsch. Math.-Ver. **121**, 71–90 (2019)

Chapter 2
Prescribed Energy Periodic Solutions of Kepler Problems with Relativistic Corrections

Alberto Boscaggin, Walter Dambrosio, and Guglielmo Feltrin

MSC Classification 34C25, 70H08, 70H12, 83A05, 83C10

2.1 Introduction

The Kepler problem of classical mechanics consists of the study of the motion of a nonattracting body of mass m in the gravitational field of a second fixed body. According to Newton's law, it can thus be described by the equation

$$m\ddot{x} = -\alpha \frac{x}{|x|^3}, \qquad x \in \mathbb{R}^d \setminus \{0\}, \tag{2.1}$$

where $\alpha = GMm$, with G the gravitational constant and M the mass of the second body (placed at the origin, for simplicity), $d \in \mathbb{N} \setminus \{0\}$, see for instance [11, 21].

With the aim of including relativistic effects in the above problem, some variants of Eq. (2.1) have been proposed. A first one, theoretically framed in the context of special relativity, comes from the use of the relativistic kinetic energy

$$K(\dot{x}) = mc^2\big(1 - \sqrt{1 - |\dot{x}|^2/c^2}\big)$$

instead of the usual one, where $c > 0$ is the speed of light (notice that $K(\dot{x}) = \frac{1}{2}m|\dot{x}|^2 + o(c^{-2})$ for $c \to +\infty$); with no modifications for the Newtonian gravitational energy, the corresponding Euler–Lagrange equation thus reads as

A. Boscaggin · W. Dambrosio (✉)
Department of Mathematics "Giuseppe Peano", University of Torino, Torino, Italy
e-mail: alberto.boscaggin@unito.it; walter.dambrosio@unito.it

G. Feltrin
Department of Mathematics, Computer Science and Physics, University of Udine, Udine, Italy
e-mail: guglielmo.feltrin@uniud.it

P. Amster, P. Benevieri (eds.), *Topological Methods for Delay and Ordinary Differential Equations*, Advances in Mechanics and Mathematics 51,
https://doi.org/10.1007/978-3-031-61337-1_2

$$\frac{\mathrm{d}}{\mathrm{d}t}\left(\frac{m\dot{x}}{\sqrt{1-|\dot{x}|^2/c^2}}\right) = -\alpha\,\frac{x}{|x|^3}, \qquad x \in \mathbb{R}^d \setminus \{0\}, \tag{2.2}$$

see, among others, [2, 9, 14, 16, 19]. A second relativistic variant of Eq. (2.1) arises, instead, by keeping the usual kinetic energy but imposing a correction for the gravitational potential: Precisely

$$V(x) = -\frac{\kappa(c)}{|x|} - \frac{\lambda(c)}{|x|^2},$$

where $\kappa(c)$ and $\lambda(c)$ are suitable positive constants with the property that $\kappa(c) \to \alpha$ and $\lambda(c) \to 0$ as $c \to +\infty$. The reason for this ad hoc correction of the Newtonian potential, suggested at first by Levi-Civita [17], lies in the fact that the corresponding equation

$$m\ddot{x} = -\kappa(c)\frac{x}{|x|^3} - 2\lambda(c)\frac{x}{|x|^4}, \qquad x \in \mathbb{R}^d \setminus \{0\}, \tag{2.3}$$

turns out to be consistent with the theory of general relativity, see for instance [13, Chapter 12.3 B].

Besides being well recognized in the physics community, Eqs. (2.2) and (2.3) have attracted the attention of mathematicians from dynamical systems and nonlinear analysis: The main problem in this context is to understand which aspects of the unperturbed dynamics are preserved if some coefficient is allowed to vary with time and/or if some external force is assumed to act on the body in addition to the Newtonian gravity. Contributions in this direction dealing with Eq. (2.3) are more classical and can be found, among others, in [1, 12, 15, 20]. On the other hand, probably due to the nonlinear nature of the involved differential operator [18], the interest for Eq. (2.2) seems to be more recent, and we can quote the papers [5, 7, 8, 22, 26].

Following this line of research, in this chapter, we consider, in dimension $d = 2$, the following perturbed versions of the above equations,

$$\frac{\mathrm{d}}{\mathrm{d}t}\left(\frac{m\dot{x}}{\sqrt{1-|\dot{x}|^2/c^2}}\right) = -\alpha\,\frac{x}{|x|^3} + \varepsilon\,\nabla U(x), \qquad x \in \mathbb{R}^2 \setminus \{0\}, \tag{2.4}$$

and

$$m\ddot{x} = -\kappa(c)\frac{x}{|x|^3} - 2\lambda(c)\frac{x}{|x|^4} + \varepsilon\,\nabla U(x), \qquad x \in \mathbb{R}^2 \setminus \{0\}, \tag{2.5}$$

where ε is a small real parameter and U is a C^∞ external potential. In this setting, our interest is in investigating the existence of periodic solutions bifurcating from noncircular periodic solutions of the corresponding unperturbed problem ($\varepsilon = 0$). Let us notice that a strictly related problem was recently considered in [5, 6]:

However, while therein the bifurcating solutions were required to have *prescribed period* $T > 0$, here we focus on the complementary, and more delicate, case of solutions with *prescribed energy* $H \in \mathbb{R}$. This means that

$$\frac{1}{2}m|\dot{x}(t)|^2 - \frac{\kappa(c)}{|x(t)|} - \frac{\lambda(c)}{|x(t)|^2} - \varepsilon\, U(x(t)) = H,$$

when $x(t)$ is a solution of the more classical (2.5); on the other hand, in view of the Lagrangian structure of (2.4), it means that

$$mc^2\left(\frac{1}{\sqrt{1-|\dot{x}|^2/c^2}} - 1\right) - \frac{\alpha}{|x|} - \varepsilon\, U(x) = H,$$

whenever $x(t)$ solves (2.4). Notice that, in principle, the values of H carrying noncircular periodic solutions are not known and must be characterized by phase-plane analysis arguments.

After having performed this preliminary step, the main tool for the proof of our result is an abstract bifurcation theory from periodic manifolds, developed by A. Weinstein in a series of papers [23–25]. Roughly speaking, this theory ensures that whenever there is a nondegenerate manifold Σ of periodic solutions with energy H for a Hamiltonian system $\dot{z} = X_{\mathcal{K}_0}(z)$ on a symplectic manifold M, then a finite number of periodic solutions, with energy $\mathcal{H}_\varepsilon = H$, exist for the system

$$\dot{z} = X_{\mathcal{H}_\varepsilon}(z)$$

whenever the perturbed Hamiltonian $\mathcal{H}_\varepsilon$ is close enough to the unperturbed one $\mathcal{K}_0$. We point out that the notion of nondegeneracy, in this fixed-energy context, is quite delicate, and several nonequivalent notions have been considered in the literature, see Remark 2.1.

In our paper, Weinstein's theory is combined with tools from Hamiltonian dynamics. Indeed, both Eqs. (2.2) and (2.3) can be regarded as Hamiltonian systems with two degrees of freedom x and p, where the momentum p is given by $p = m\dot{x}$ in the more classical case of Eq. (2.3), while $p = m\dot{x}/\sqrt{1-|\dot{x}|^2/c^2}$ for Eq. (2.2). As a consequence of the invariances by time translation and space rotations, two independent integrals can be found: the energy and the angular momentum $x \wedge p$. Thus, assuming that a periodic solution x with energy H is found, the periodic manifold Σ for the application of the abstract bifurcation theory is nothing but the two-dimensional torus containing x together with all its time translations and space rotations. After passing to action-angle coordinates, the nondegeneracy condition becomes the standard nondegeneracy condition used in the isoenergetic version of KAM theory, see for instance [3, 10]. By explicit computations, such a condition is easily checked to hold true both in the case of (2.2) and (2.3), thus concluding the proof.

The plan of the chapter is the following. In Sect. 2.2, we review Weinstein's theory, with a focus to the case of nearly integrable Hamiltonian systems, needed for our application. Then, in Sect. 2.3 we give the statement and the proof of our main results for Eqs. (2.4) and (2.5).

2.2 Preliminaries

In this section, we state and prove a result ensuring, for a nearly integrable Hamiltonian system, bifurcation of prescribed energy periodic solutions from periodic tori of the integrable unperturbed problem. This result is obtained as a corollary of a general bifurcation theory from periodic manifolds by A. Weinstein [23–25]. For the reader's convenience, we first briefly review this theory in Sect. 2.2.1, and then in Sect. 2.2.2 we provide our result (Theorem 2.2).

2.2.1 *An Abstract Bifurcation Result by Weinstein*

Let (M, ω) be a symplectic manifold. Given $\mathcal{K}_0 \in C^\infty(M)$ we consider the Hamiltonian system

$$\dot{z} = X_{\mathcal{K}_0}(z), \tag{2.6}$$

where as usual $X_{\mathcal{K}_0}$ denotes the Hamiltonian vector field associated to $\mathcal{K}_0$ (that is, $\omega(X_{\mathcal{K}_0}, Y) = d\mathcal{K}_0(Y)$ for every vector field Y on M). For every $\xi \in M$, the unique (local) solution of (2.6) satisfying $z(0) = \xi$ will be denoted by $z(t; \xi)$; as well-known, the value of the Hamiltonian $\mathcal{K}_0$ is constant along $z(t; \xi)$, that is

$$\mathcal{K}_0(z(t; \xi)) = H, \qquad \text{for every } t,$$

where $H = \mathcal{K}_0(\xi)$.

We are interested in values H such that the level set $\mathcal{K}_0^{-1}(H)$ contains nonconstant periodic solutions of (2.6). More precisely, let us define the set $\mathrm{Per}^H_{\mathcal{K}_0}$ of periodic points with energy H as the set of couples $(\xi, \tau) \in M \times (0, +\infty)$ such that $z(t; \xi)$ is a nonconstant τ-periodic solution of (2.6) with energy H, that is,

$$z(\tau; \xi) = \xi, \qquad \mathcal{K}_0(\xi) = H, \qquad X_{\mathcal{K}_0}(\xi) \neq 0.$$

Incidentally, notice that τ does not need to be the minimal period, and, actually, if (ξ, τ) belongs to $\mathrm{Per}^H_{\mathcal{K}_0}$, then $(\xi, N\tau)$ belongs to $\mathrm{Per}^H_{\mathcal{K}_0}$ as well, for every integer $N \geq 1$. Moreover, of course, if (ξ, τ) belongs to $\mathrm{Per}^H_{\mathcal{K}_0}$, then the one-dimensional torus $\{(z(t; \xi), \tau) : t \in [0, \tau)\}$ belongs to $\mathrm{Per}^H_{\mathcal{K}_0}$ as well.

We say that a set $\Sigma \subset \mathrm{Per}^H_{\mathcal{K}_0}$ is a *periodic manifold* if it satisfies the following conditions:

(i) Σ is a closed submanifold of $M \times \mathbb{R}$.

(ii) The restriction to Σ of the projection $\pi : M \times \mathbb{R} \to M$ is an embedding.

Notice that, roughly, condition (ii) ensures that for a periodic point $(\xi, \tau) \in \Sigma$, the period has to be uniquely chosen (though not necessarily the minimal one and, in general, depending, smoothly, on ξ).

The crucial additional assumption to be imposed in order to ensure bifurcation of periodic solutions (with energy H) from the periodic manifold Σ is a nondegeneracy condition. To introduce it, we consider, for any $(\xi, \tau) \in \Sigma$, the so-called monodromy operator at ξ, that is, the linear operator $P : T_\xi M \to T_\xi M$ given by

$$P\eta = \partial_\zeta z(\tau; \zeta)|_{\zeta=\xi}\, \eta.$$

It is well-known that both $X_{\mathcal{K}_0}(\xi)$ and $T_\xi(\mathcal{K}_0^{-1}(H))$ are invariant under P; moreover, $X_{\mathcal{K}_0}(\xi) \in T_\xi(\mathcal{K}_0^{-1}(H))$ (incidentally, let us observe that, since $X_{\mathcal{K}_0}(\xi) \neq 0$, the level set $\mathcal{K}_0^{-1}(H)$ is a submanifold around ξ and thus the tangent space $T_\xi(\mathcal{K}_0^{-1}(H))$ is well-defined). With this in mind, a periodic manifold Σ is said to be *nondegenerate* if the following condition holds:

(iii) For every $(\xi, \tau) \in \Sigma$ and for every $\eta \in T_\xi(\mathcal{K}_0^{-1}(H))$ it holds that

$$\eta \in T_\xi \pi(\Sigma) \quad \Longleftrightarrow \quad \eta = P\eta + \lambda X_{\mathcal{K}_0}(\xi), \quad \text{for some } \lambda \in \mathbb{R}$$

(cf. [23, Lemma 1.1]).

Remark 2.1 The above nondegeneracy condition might seem quite mysterious at first sight, since in principle one could expect the eigenspace $\mathcal{E}' \subset T_\xi(\mathcal{K}_0^{-1}(H))$ of fixed points of P to play a role. The reason why the larger space

$$\mathcal{E} = \left\{ \eta \in T_\xi(\mathcal{K}_0^{-1}(H)) : \eta = P\eta + \lambda X_{\mathcal{K}_0}(\xi), \text{ for some } \lambda \in \mathbb{R} \right\} \supset \mathcal{E}'$$

has to be considered lies in the fact that we are dealing with periodic solutions with *unprescribed period*. Indeed, if $(\xi(s), \tau(s))$ is a smooth curve in Σ with $(\xi(0), \tau(0)) = (\xi, \tau)$, then differentiating with respect to s and taking $s = 0$ yield

$$\xi'(0) = P\xi'(0) + \tau'(0) X_{\mathcal{K}_0}(\xi(0)),$$

where, in general, $\tau'(0)$ does not need to be zero. This shows that

$$T_\xi \pi(\Sigma) \subset \mathcal{E},$$

and thus the nondegeneracy conditions amount in requiring that all vectors in $\mathcal{E}$ belong to $T_\xi \pi(\Sigma)$. We stress that, on the other hand, the fact that periodic solutions are required to have *prescribed energy* leads to consider the monodromy operator P not on the whole $T_\xi M$, but on its invariant subspace $T_\xi(\mathcal{K}_0^{-1}(H))$.

We point out, however, that other nondegeneracy conditions for the periodic problem with prescribed energy have been considered in the literature. For instance, in [24] the additional condition that the algebraic multiplicity of 1 as an eigenvalue of P is equal to the dimension of Σ is imposed. As we will see in Remark 2.3, this stronger assumption is not satisfied in the applications which we are going to develop in Sect. 2.2.2. $\triangleleft$

We are now in a position to state the result about bifurcation of periodic solutions (with prescribed energy) from a nondegenerate periodic manifold of system (2.6); it corresponds to [23, Theorem 1.4] in the case of the action of the group $\mathbb{Z}_n$ for $n = 1$.

Theorem 2.1 *Let (M, ω) be a symplectic manifold such that the form ω is exact, let $\mathcal{K}_0 \in C^\infty(M)$, and let $\Sigma \subset \mathrm{Per}^H_{\mathcal{K}_0}$ a nondegenerate periodic manifold, for some $H \in \mathbb{R}$. Finally, consider a function $\mathcal{H} \in C^\infty(M \times \mathbb{R})$ such that $\mathcal{H}_0 = \mathcal{K}_0$, where $\mathcal{H}_\varepsilon = \mathcal{H}(\cdot, \varepsilon)$. Then, denoting by m the least integer greater than or equal to* $\mathrm{Cat}(\Sigma)/2$, *for every neighborhood $\mathcal{U} \subset M \times (0, +\infty)$ of Σ, there exists $\varepsilon^* > 0$ such that if $|\varepsilon| < \varepsilon^*$, there exist $\tau_1, \ldots, \tau_m > 0$ and there exist $z_1, \ldots, z_m$ distinct periodic solutions of the Hamiltonian system*

$$\dot{z} = X_{\mathcal{H}_\varepsilon}(z)$$

such that for every $i = 1, \ldots, m$:

(I) $\mathcal{H}_\varepsilon(z_i(t)) = H$, *for every* $t \in \mathbb{R}$.
(II) τ_i *is a period for* z_i.
(III) $\{(z_i(t), \tau_i) : t \in \mathbb{R}\} \subset \mathcal{U}$.

Notice that (III) provides a localization information for both the orbit $\{z_i(t)\}_{t\in\mathbb{R}}$ on M and the period τ_i, which can thus be chosen arbitrarily near orbits and periods of the solutions of the unperturbed problem. We also point out that, actually, the exactness condition on the symplectic form ω can be removed, as shown later in [25].

2.2.2 Bifurcation from Periodic Invariant Tori

Let $\mathcal{D} \subset \mathbb{R}^n$ $(n \geq 2)$ be an open set and let $\mathcal{K}_0 : \mathcal{D} \to \mathbb{R}$ and $\mathcal{R} : \mathcal{D} \times \mathbb{R}^n \to \mathbb{R}$ be C^∞ functions; moreover, we assume that $\mathcal{R}$ is 2π-periodic in each variable φ_i, where $\varphi = (\varphi_1, \ldots, \varphi_n)$. Defining the symplectic form

$$\omega = \mathrm{d}\varphi \wedge \mathrm{d}I \tag{2.7}$$

and, for every $\varepsilon \in \mathbb{R}$, the function $\mathcal{H}_\varepsilon : \mathcal{D} \times \mathbb{R}^n \to \mathbb{R}$ as

$$\mathcal{H}_\varepsilon(I, \varphi) = \mathcal{K}_0(I) + \varepsilon\, \mathcal{R}(I, \varphi),$$

the Hamiltonian system associated with the Hamiltonian $\mathcal{H}_\varepsilon$ with respect to the symplectic structure induced by the form ω reads as

$$\begin{cases} \dot{I} = -\varepsilon \nabla_\varphi \mathcal{R}(I, \varphi), \\ \dot{\varphi} = \nabla \mathcal{K}_0(I) + \varepsilon \nabla_I \mathcal{R}(I, \varphi). \end{cases} \tag{2.8}$$

For $\varepsilon = 0$, system (2.8) reduces to the completely integrable system in action-angle coordinates

$$\begin{cases} \dot{I} = 0, \\ \dot{\varphi} = \nabla \mathcal{K}_0(I), \end{cases} \tag{2.9}$$

whose solutions are given by

$$I(t) = I^*, \quad \varphi(t) = \varphi^* + t\,\nabla \mathcal{K}_0(I^*), \quad t \in \mathbb{R}, \tag{2.10}$$

being $(I^*, \varphi^*) \in \mathcal{D} \times \mathbb{R}^n$ the initial datum.

In the following, (2.8) is meant as the lifting to the covering space of the corresponding Hamiltonian system on the manifold

$$M = \mathcal{D} \times \mathbb{T}^n, \tag{2.11}$$

where $\mathbb{T}^n = \mathbb{R}^n / 2\pi\,\mathbb{Z}^n$ is the n-dimensional torus. Accordingly, we say that a solution of (2.8) is periodic if its projection on M is periodic, that is, if there exists $\tau^* > 0$ such that

$$I(\tau^*) = I(0), \qquad \varphi(\tau^*) - \varphi(0) \in 2\pi\,\mathbb{Z}^n.$$

Hence, if $I^* \in \mathcal{D}$ and $\tau^* > 0$ are such that

$$\tau^* \nabla \mathcal{K}_0(I^*) \in 2\pi\,\mathbb{Z}^n \setminus \{0\}, \tag{2.12}$$

and if τ^* is the least positive number with the above property, then the torus

$$\mathcal{T}_{I^*} = \{I^*\} \times \mathbb{T}^n \tag{2.13}$$

is filled by nonconstant periodic solutions of system (2.9) of minimal period τ^*.

In this setting, we are interested in the existence of periodic solutions to (2.8), having prescribed energy, bifurcating from the torus $\mathcal{T}_{I^*}$. Our result is the following.

Theorem 2.2 *Let us assume that condition* (2.12) *holds true and that*

$$\det \begin{pmatrix} \nabla^2 \mathcal{K}_0(I^*) & \nabla \mathcal{K}_0(I^*)^\top \\ \nabla \mathcal{K}_0(I^*) & 0 \end{pmatrix} \neq 0, \tag{2.14}$$

where $\nabla\mathcal{K}_0(I^*)$ is meant as a row vector. Let $H^* = \mathcal{K}_0(I^*)$ and let m be the least integer greater than or equal to $(n+1)/2$. Then, for every $\varsigma > 0$, there exists $\varepsilon^* > 0$ such that, if $|\varepsilon| \leq \varepsilon^*$, there exist $\tau_1, \ldots, \tau_m > 0$ and there exist $(I_1, \varphi_1), \ldots, (I_m, \varphi_m)$ distinct periodic solutions of the Hamiltonian system (2.8) such that, for every $i = 1, \ldots, m$:

(I) $\mathcal{K}_\varepsilon(I_i(t), \varphi_i(t)) = H^*$, *for every* $t \in \mathbb{R}$.
(II) τ_i *is the minimal period for* (I_i, φ_i).
(III) The conditions

$$|\tau_i - \tau^*| < \varsigma \tag{2.15}$$

and

$$|I_i(t) - I^*| < \varsigma, \quad |\varphi_i(t) - \varphi_i(0) - t\,\nabla\mathcal{K}_0(I^*)| < C_*\varsigma, \quad \forall\, t \in [0, \tau_i], \tag{2.16}$$

are satisfied, where $C_* > 0$ is a constant depending only on $\mathcal{K}_0$ and $\mathcal{R}$.

Proof Regarding (2.8) as a Hamiltonian system on the symplectic manifold (M, ω), with M and ω as in (2.11) and (2.7), respectively, and noticing that ω is exact on M, we are going to apply Theorem 2.1 with the choice

$$\Sigma = \mathcal{T}_{I^*} \times \{\tau^*\} \subset \mathrm{Per}^{H^*}_{\mathcal{K}_0},$$

with $\mathcal{T}_{I^*}$ as in (2.13). Of course, Σ is diffeomorphic to a n-dimensional torus, so that condition (i) holds true; moreover, since the period is constantly equal to τ^* for every solution on $\mathcal{T}_{I^*}$, condition (ii) is trivially satisfied.

It thus remains to check that the nondegeneracy condition (iii) is satisfied. So, let us fix $(\xi, \tau) \in \Sigma$, that is $\tau = \tau^*$ and $\xi = (I^*, \varphi)$ for some $\varphi \in [0, 2\pi)$. Then,

$$T_\xi(\mathcal{K}_0^{-1}(H^*)) = \left\{\eta = (v, \phi) \in \mathbb{R}^n \times \mathbb{R}^n \colon \langle \nabla\mathcal{K}_0(I^*), v\rangle = 0\right\}$$

and, of course,

$$T_\xi \pi(\Sigma) = \{0\} \times \mathbb{R}^n.$$

On the other hand,

$$X_{\mathcal{K}_0}(\xi) = (0, \nabla\mathcal{K}_0(I^*)),$$

while recalling (2.10) we easily infer that the monodromy operator P is given by

$$P\eta = \begin{pmatrix} \mathrm{Id}_{\mathbb{R}^n} & 0 \\ \tau^*\nabla^2\mathcal{K}_0(I^*) & \mathrm{Id}_{\mathbb{R}^n} \end{pmatrix} \eta = (v, \tau^*\nabla^2\mathcal{K}_0(I^*)v + \phi), \tag{2.17}$$

for every $\eta = (v, \phi) \in T_\xi(\mathcal{K}_0^{-1}(H^*))$. Hence

$$(\mathrm{Id}_{\mathbb{R}^{2n}} - P)\eta = (0, \tau^*\nabla^2\mathcal{K}_0(I^*)v)$$

and thus condition (iii) is satisfied if and only if, given $\lambda \in \mathbb{R}$, the only vector $v \in \mathbb{R}^n$ satisfying

$$\tau^*\nabla^2\mathcal{K}_0(I^*)v = \lambda\nabla\mathcal{K}_0(I^*)$$

and such that $\langle\nabla\mathcal{K}_0(I^*), v\rangle = 0$ is the vector $v = 0$.

Let us assume by contradiction that this is not case. Hence, setting $\lambda' = -\lambda/\tau^*$ we have found a nontrivial solution (v, λ') of the linear homogeneous system in $\mathbb{R}^{n+1}$ whose coefficient matrix is

$$\begin{pmatrix} \nabla^2\mathcal{K}_0(I^*) & \nabla\mathcal{K}_0(I^*)^\intercal \\ \nabla\mathcal{K}_0(I^*) & 0 \end{pmatrix},$$

thus contradicting assumption (2.14).

Summing up, all the assumptions of Theorem 2.1 are satisfied. Hence, denoting by m the least integer greater than or equal to $\mathrm{Cat}(\Sigma)/2 = \mathrm{Cat}(\mathbb{T}^n)/2 = (n+1)/2$, given any $\varsigma > 0$, we can consider the open neighborhood U of Σ given by

$$U = B_\varsigma(I^*) \times \mathbb{T}^n \times (\tau^* - \varsigma, \tau^* + \varsigma)$$

and apply Theorem 2.1 to find, for $|\varepsilon| \leq \varepsilon^*$, m numbers $\tau_i > 0$ satisfying (2.15) and m solutions (I_i, φ_i) of system (2.8), with period τ_i, energy H and satisfying the first condition in (2.16). We now claim that condition (2.16) is satisfied, as well. To prove this, we assume without loss of generality that $\varepsilon^* \leq \varsigma$, and we first notice that the first condition in (2.16), together with the regularity of $\mathcal{K}_0$ and $\mathcal{R}$, implies that

$$|\nabla\mathcal{K}_0(I_i(t)) - \nabla\mathcal{K}_0(I^*)| < \frac{C_*}{2\tau_i}\varsigma, \quad \text{for every } t \in [0, \tau_i],$$

and that

$$|\nabla_I\mathcal{R}(I_i(t), \varphi_i(t))| < \frac{C_*}{2\tau_i}, \quad \text{for every } t \in [0, \tau_i],$$

where $C_* > 0$ is a constant depending only on $\mathcal{K}_0$ and $\mathcal{R}$. Since

$$\begin{aligned}\varphi_i(t) - \varphi_i(0) - t\,\nabla\mathcal{K}_0(I^*) &= \\ &= \int_0^t \left(\dot\varphi_i(s) - \nabla\mathcal{K}_0(I^*)\right) \mathrm{d}s\end{aligned}$$

$$= \int_0^t \left(\nabla\mathcal{K}_0(I_i(s)) - \nabla\mathcal{K}_0(I^*)\right) \mathrm{d}s + \varepsilon \int_0^t \nabla_I \mathcal{R}(I_i(s), \varphi_i(s))\, \mathrm{d}s,$$

we thus have

$$\left|\varphi_i(t) - \varphi_i(0) - t\,\nabla\mathcal{K}_0(I^*)\right| \leq C_* \varsigma, \quad \text{for every } t \in [0, \tau_i],$$

as desired.

It thus remains to show that τ_i is the minimal period of (I_i, φ_i). To this end, we first observe that (2.15) and (2.16) imply, for ς small enough, that

$$\varphi_i(\tau_i) - \varphi_i(0) = \tau^* \nabla\mathcal{K}_0(I^*).$$

Now, let us assume by contradiction that the period τ_i is not minimal, that is, τ_i/ℓ is also a period for some integer $\ell \geq 2$. Hence, both the components of the vector $\varphi_i(\tau_i) - \varphi_i(0)$ must be a multiple of ℓ, and thus

$$\frac{\tau^*}{\ell} \nabla\mathcal{K}_0(I^*) \in 2\pi\, \mathbb{Z}^n \setminus \{0\}.$$

This contradicts that fact that τ^* is the least number satisfying (2.12) (that is to say, τ^* is not the minimal period of the solution on the unperturbed torus $\mathcal{T}_{I^*}$). □

Remark 2.2 In Sect. 2.3 we will apply Theorem 2.2 in the case of two degrees of freedom, that is, $n = 2$. It is easy to see that in this situation the nondegeneracy condition (2.14) writes equivalently as

$$\langle \nabla^2\mathcal{K}_0(I^*)(\partial_2\mathcal{K}_0(I^*), -\partial_1\mathcal{K}_0(I^*)), (\partial_2\mathcal{K}_0(I^*), -\partial_1\mathcal{K}_0(I^*))\rangle \neq 0. \tag{2.18}$$

We also explicitly observe that the number of bifurcating periodic solutions of (2.8) is 2. ◁

Remark 2.3 From (2.17), it is immediate to deduce that 1 is the only eigenvalue of P: Thus, its algebraic multiplicity is $2n - 1$, a number that is (for $n \geq 2$) strictly greater than n, that is, the dimension of the manifold Σ. Therefore, the nondegeneracy condition of [24] is not satisfied in our setting, cf. Remark 2.1. ◁

2.3 Main Results

In this section, we apply Theorem 2.2 to study bifurcation of periodic solutions for the two relativistic problems described in the Introduction: the Kepler problem with Levi-Civita correction (2.5) and the Kepler problem with relativistic differential operator (2.4).

The two problems share some structural features that enable us to introduce a common strategy to deal with. Indeed, they both have a Hamiltonian formulation, and the unperturbed problems are integrable, due to the presence of two independent first integrals in involution (the energy and the angular momentum).

The construction of action-angle coordinates in both the cases is based on a phase-plane analysis for the radial component of the solutions in polar coordinates.

For the purpose of applying Theorem 2.2, we first characterize the pairs energy/angular momentum for which periodic solutions exist, and then we directly prove the validity of the nondegeneracy condition (2.18).

2.3.1 The Kepler Problem with Levi-Civita Correction

In this section we deal with the problem

$$m\ddot{x} = -\kappa \frac{x}{|x|^3} - 2\lambda \frac{x}{|x|^4} + \varepsilon \nabla U(x), \tag{2.19}$$

where $m > 0$, $\kappa > 0$, $\lambda > 0$, $U \in C^\infty(\mathbb{R}^2, \mathbb{R})$, and $\varepsilon \in \mathbb{R}$.

Our result is the following (by now, we do not give any physical interpretation for the constants κ and λ: This aspect will be treated in the discussion leading to Corollary 2.1).

Theorem 2.3 *Let $H^* < 0$ and $N \in \mathbb{N}$. Then, for every $\varsigma > 0$ there exists $\varepsilon^* = \varepsilon^*(N, \varsigma) > 0$ such that for every $\varepsilon \in \mathbb{R}$, with $|\varepsilon| \le \varepsilon^*$, Eq. (2.19) has at least $2N$ periodic solutions $x_1^i, \dots, x_N^i$, $i = 1, 2$, of energy H^*.*

Moreover, for every $j = 1, \dots, N$ and $i = 1, 2$, the solution x_j^i has minimal period $T_{i,j}$ satisfying

$$\left| T_{i,j} - \frac{\pi \kappa \sqrt{m}}{\sqrt{2}(-H^*)^{3/2}} \right| < \varsigma$$

and winding number in its minimal period

$$\left\lfloor \frac{\sqrt{-2H^*}}{\kappa} \sqrt{2\lambda + \frac{\kappa^2}{-2H^*}} \right\rfloor + j,$$

where $\lfloor \cdot \rfloor$ denotes the integer part.

Theorem 2.3 follows from the application of Theorem 2.2 to the Hamiltonian system associated to (2.19). Indeed, let us first recall that (2.19) can be written as a Hamiltonian system, with respect to the variables $(x, p) = (x, m\dot{x})$. The Hamiltonian is given by

$$\mathcal{H}_\varepsilon(x,p) = \frac{1}{2m}|p|^2 - \frac{\kappa}{|x|} - \frac{\lambda}{|x|^2} - \varepsilon\, U(x),$$

whose values correspond to values of the energy

$$E_\varepsilon(x,\dot{x}) = \frac{1}{2}m|\dot{x}|^2 - \frac{\kappa}{|x|} - \frac{\lambda}{|x|^2} - \varepsilon\, U(x). \tag{2.20}$$

Now, let us observe that the Hamiltonian system associated to $\mathcal{H}_\varepsilon$ is nearly integrable and that can be transformed in the form (2.8) passing to action-angle coordinates. The construction is quite standard (cf. [6, Section 2]), and it relies on the fact that the unperturbed Hamiltonian system associated to $\mathcal{H}_0$ has a second first integral independent from $\mathcal{H}_0$ and in involution with it, which is the angular momentum defined by

$$\mathcal{L}_0(x,p) = \langle x, Jp\rangle, \qquad \text{where } J = \begin{pmatrix} 0 & 1 \\ -1 & 0 \end{pmatrix}. \tag{2.21}$$

In order to define the change of variable, we focus on the unperturbed problem

$$m\ddot{x} = -\kappa\frac{x}{|x|^3} - 2\lambda\frac{x}{|x|^4}, \tag{2.22}$$

and we pass to polar coordinates $x = re^{i\vartheta}$.

Let us define

$$\Lambda = \left\{(H,L) \in \mathbb{R}^2 \colon L^2 \in \left(\frac{2\lambda}{m}, +\infty\right),\ H \in \left(-\frac{\kappa^2}{2(mL^2-2\lambda)}, 0\right)\right\}.$$

Following [6], we know that the radial component r of the solutions of (2.22) satisfies

$$\frac{1}{2}m\dot{r}^2 + \frac{mL^2}{2r^2} - \frac{\kappa}{r} - \frac{\lambda}{r^2} = H.$$

From the above relation we infer that for every $L \in \mathbb{R}$ the radial component moves accordingly to a nonlinear oscillator driven by the effective potential

$$W(r;L) = \frac{mL^2 - 2\lambda}{2r^2} - \frac{\kappa}{r}.$$

When $mL^2 > 2\lambda$, this potential satisfies

$$\lim_{r\to 0^+} W(r;L) = +\infty, \qquad \lim_{r\to+\infty} W(r;L) = 0,$$

and it is strictly decreasing in $(0, (mL^2 - 2\lambda)/\kappa)$ and strictly increasing in $((mL^2 - 2\lambda)/\kappa, +\infty)$. The global minimum of $W(\cdot; L)$ is

$$w_{\min}(L) = -\frac{\kappa^2}{2(mL^2 - 2\lambda)}. \tag{2.23}$$

As a consequence, for every $(H, L) \in \Lambda$ the orbits in the $(r, \dot{r})$-plane are closed curves.

In [6, Section 4.1] it is proved that the corresponding radial period is given by

$$T(H, L) = \frac{\pi\kappa\sqrt{m}}{\sqrt{2}(-H)^{3/2}} \tag{2.24}$$

and that the angular displacement in this period (the so-called *apsidal angle*) is

$$\Theta(H, L) = \frac{2\pi\sqrt{m}L}{\sqrt{mL^2 - 2\lambda}}.$$

For every $(H, L) \in \Lambda$, the set

$$\mathcal{T}_{(H,L)} = \left\{(x, p) \in (\mathbb{R}^2 \setminus \{0\}) \times \mathbb{R}^2 : \mathcal{H}_0(x, p) = H,\ \mathcal{L}_0(x, p) = L\right\}$$

is then diffeomorphic to a two-dimensional torus $\mathbb{T}^2$ (cf. [3, 4]).

Action-angle coordinates can be constructed in a neighborhood of $\mathcal{T}_{(H^*,L^*)}$, for every fixed $(H^*, L^*) \in \Lambda$. Indeed, given $(H^*, L^*) \in \Lambda$, for every (H, L) in a neighborhood Υ^* of (H^*, L^*), let $\mathcal{A}(H, L)$ be the area of the bounded region enclosed by the orbit corresponding to (H, L) in the $(r, \dot{r})$-plane and define $\Pi\colon \Upsilon^* \to \mathbb{R}^4$ by

$$\Pi(H, L) = \left(\frac{1}{2\pi}\mathcal{A}(H, L) + L, L, 2\pi\frac{\mu}{T(H, L)}, (\Theta(H, L) - 2\pi)\frac{\mu}{T(H, L)} + \psi\right), \tag{2.25}$$

for every $(H, L) \in \Upsilon^*$, where $\mu \in [0, T(H, L))$ is the time needed to reach the point $(r, \dot{r})$ in the $(r, \dot{r})$-plane along the orbit starting from the pericenter and $\psi \in [0, 2\pi)$ is the angle, measured in the counter-clockwise sense from $\vartheta = 0$, of the greatest nonpositive instant in which x lies at the pericenter.

The map

$$\Xi\colon (x, p) \mapsto \Pi(\mathcal{H}_0(x, p), \mathcal{L}_0(x, p))$$

is well-defined for (x, p) in a neighborhood of the fixed torus $\mathcal{T}_{(H^*,L^*)}$, and it is well-known that it provides a symplectic diffeomorphism from this set onto its image.

In the new variables, the Hamiltonian system is of the form (2.8) with

$$\mathcal{K}_0(I_1, I_2) = \mathcal{H}_0(\Xi^{-1}(I_1, I_2, \varphi_1, \varphi_2)). \tag{2.26}$$

We are now in a position to prove Theorem 2.3.

Proof of Theorem 2.3 From the previous discussion, we know that in the action-angle variables $(I_1, I_2, \varphi_1, \varphi_2)$ the Hamiltonian system corresponding to (2.19) is a smooth perturbation of the Hamiltonian system associated to the Hamiltonian $\mathcal{K}_0$ defined in (2.26).

Now, let us fix $H^* < 0$, $N \in \mathbb{N}$, $j \in \{1, \dots N\}$ and let us set

$$k_j^* = \left\lfloor \frac{\sqrt{-2H^*}}{\kappa} \sqrt{2\lambda + \frac{\kappa^2}{-2H^*}} \right\rfloor + j.$$

Let

$$L^*_{\max} = \frac{1}{\sqrt{m}} \sqrt{2\lambda + \frac{\kappa^2}{-2H^*}}.$$

Notice that $L^*_{\max}$ is the unique solution of

$$H^* = -\frac{\kappa^2}{2(mL^2 - 2\lambda)}$$

and that $(H^*, L^*) \in \Lambda$, for every $L^* \in (\sqrt{2\lambda/m}, L^*_{\max})$. Moreover, since

$$2\pi \frac{\sqrt{-2H^*}}{\kappa} \sqrt{2\lambda + \frac{\kappa^2}{-2H^*}} = \Theta(H^*, L^*_{\max}),$$

from the monotonicity of $\Theta(H^*, \cdot)$ and the choice of k_j^*, we deduce that there exists a unique $L_j^* \in (\sqrt{2\lambda/m}, L^*_{\max})$ such that

$$\Theta(H^*, L_j^*) = 2\pi\, k_j^*. \tag{2.27}$$

Since $(H^*, L_j^*) \in \Lambda$, as already observed, we deduce that the solution of the unperturbed equation (2.22) having energy H^* and angular momentum L_j^* is periodic of period $T(H^*, L_j^*)$, with T as in (2.24), and it has winding number $2\pi k_j^*$ in a period.

As far as (2.18) is concerned, with $\mathcal{K}_0$ given in (2.26), we prove its validity for every (I_1, I_2) in the domain of $\mathcal{K}_0$, which will be denoted by O^*. Indeed, let us first observe that by construction of the action-angles variables, we have $\mathcal{K}_0(I_1, I_2) = H(I_1, I_2)$, where $(H, L) = (H(I_1, I_2), L(I_1, I_2))$ satisfies

$$I_1 = \frac{1}{2\pi}\mathcal{A}(H, L) + L, \qquad I_2 = L$$

(recall (2.25)). From these relations we immediately deduce that $L(I_1, I_2) = I_2$ and that $H = H(I_1, I_2)$ fulfills

$$\frac{1}{2\pi}\mathcal{A}(H, I_2) = I_1 - I_2. \tag{2.28}$$

Now, let us recall that $\mathcal{A}(H, L) = \partial_H T(H, L)$, for every $(H, L) \in \Lambda$ (see [3, p. 282]). Taking into account (2.24), we infer that

$$\mathcal{A}(H, L) = \frac{2\pi\kappa}{\sqrt{-2H}} + \zeta(L), \quad \text{for every } (H, L) \in \Lambda, \tag{2.29}$$

for some differentiable function $\zeta : (-\infty, -\sqrt{2\lambda/m}) \cup (\sqrt{2\lambda/m}, +\infty) \to \mathbb{R}$. For every $L \in (-\infty, -\sqrt{2\lambda/m}) \cup (\sqrt{2\lambda/m}, +\infty)$, the value $\zeta(L)$ can be easily obtained recalling that for the minimum admissible value of the energy $w_{\min}(L)$ (cf. (2.23)) we have

$$\lim_{H \to w_{\min}(L)} \mathcal{A}(H, L) = 0.$$

From (2.29) we then deduce that

$$\zeta(L) = -\frac{2\pi\kappa}{\sqrt{-2w_{\min}(L)}} = -2\pi\sqrt{m}\sqrt{mL^2 - 2\lambda},$$

thus concluding that

$$\mathcal{A}(H, L) = 2\pi\sqrt{m}\left(\frac{\kappa}{\sqrt{-2H}} - \sqrt{mL^2 - 2\lambda}\right), \quad \text{for every } (H, L) \in \Lambda.$$

As a consequence, (2.28) reduces to

$$\frac{\kappa\sqrt{m}}{\sqrt{-2H}} - \sqrt{m}\sqrt{mI_2^2 - 2\lambda} = I_1 - I_2.$$

By means of a standard computation, we obtain

$$\begin{aligned}\mathcal{K}_0(I_1, I_2) = H(I_1, I_2) &= -\frac{m\kappa^2}{2}\left(I_1 - I_2 + \sqrt{m}\sqrt{mI_2^2 - 2\lambda}\right)^{-2} \\ &= -\frac{m\kappa^2}{2}\gamma(I_1, I_2)^{-2}, \quad \text{for every } (I_1, I_2) \in O^*,\end{aligned}$$

where $\gamma(I_1, I_2) = I_1 - I_2 + \sqrt{m}\sqrt{mI_2^2 - 2\lambda}$. We then deduce that

$$\partial_1 \mathcal{K}_0(I_1, I_2) = m\kappa^2 \gamma(I_1, I_2)^{-3}$$
$$\partial_2 \mathcal{K}_0(I_1, I_2) = m\kappa^2 \gamma(I_1, I_2)^{-3} \left(m\sqrt{m} I_2 \left(mI_2^2 - 2\lambda \right)^{-\frac{1}{2}} - 1 \right)$$

and

$$\partial^2_{11} \mathcal{K}_0(I_1, I_2) = -3m\kappa^2 \gamma(I_1, I_2)^{-4}$$
$$\partial^2_{12} \mathcal{K}_0(I_1, I_2) = -3m\kappa^2 \gamma(I_1, I_2)^{-4} \left(m\sqrt{m} I_2 \left(mI_2^2 - 2\lambda \right)^{-\frac{1}{2}} - 1 \right)$$
$$\partial^2_{22} \mathcal{K}_0(I_1, I_2) = -3m\kappa^2 \gamma(I_1, I_2)^{-4} \left(m\sqrt{m} I_2 \left(mI_2^2 - 2\lambda \right)^{-\frac{1}{2}} - 1 \right)^2$$
$$- 2\lambda m^2 \sqrt{m} \kappa^2 \gamma(I_1, I_2)^{-3} \left(mI_2^2 - 2\lambda \right)^{-\frac{3}{2}},$$

for every $(I_1, I_2) \in \mathcal{O}^*$. Hence, we obtain

$$\det \mathcal{K}_0(I_1, I_2) = 6\lambda m^3 \sqrt{m} \kappa^4 \gamma(I_1, I_2)^{-7} \left(mI_2^2 - 2\lambda \right)^{-\frac{3}{2}} > 0,$$

for every $(I_1, I_2) \in \mathcal{O}^*$. On the other hand, since

$$\partial^2_{11} \nabla^2 \mathcal{K}_0(I_1, I_2) = -3m\kappa^2 \gamma(I_1, I_2)^{-4} < 0, \quad \text{for every } (I_1, I_2) \in \mathcal{O}^*,$$

we deduce that the trace of $\nabla^2 \mathcal{K}_0(I_1, I_2)$ is negative, for every $(I_1, I_2) \in \mathcal{O}^*$. Hence, $\nabla^2 \mathcal{K}_0(I_1, I_2)$ is definite negative, for every $(I_1, I_2) \in \mathcal{O}^*$. Condition (2.18) is then satisfied for every $I = (I_1, I_2) \in \mathcal{O}^*$.

We can now conclude. Indeed, let $(I_j^*, \varphi_j^*) = \Pi(H^*, L_j^*)$, where Π is defined in (2.25). From Theorem 2.2 we deduce that for every $\varsigma > 0$ there exists ε_j^* such that for every $\varepsilon \in \mathbb{R}$ with $|\varepsilon| \leq \varepsilon_j^*$ Eq. (2.19) has two periodic solutions x_j^i, $i = 1, 2$, with energy H^* and satisfying the required property of the minimal period. Moreover, the condition on the winding number of x_j^i follows from (2.27) taking into account (2.16) and the continuity of the winding number (cf. the argument in the proof of [5, Theorem 3.2]). The result then follows taking $\varepsilon^* = \min\{\varepsilon_j^* \colon j = 1, \ldots, N\}$. □

Remark 2.4 We mention that the isoenergetic nondegeneracy of equation (2.22) was already proved in [20] using a different set of action-angles coordinates. Our result can thus be seen as a periodic counterpart of [20] that deals with the persistence of quasi-periodic invariant tori. ◁

We end this section with a corollary of Theorem 2.3, concerning the original relativistic problem introduced by Levi-Civita in [17, Sect. 11 of Ch. 1, Sect. 9 of Ch. 2].

In the context of the relativistic approach by Levi-Civita, the parameters κ and λ are related to the speed of light c and the Einsteinian energy $E < 0$ by

$$\kappa(E, c) = mGM + \frac{4EGM}{c^2}, \quad 2\lambda(c) = \frac{6G^2M^2}{c^2}, \tag{2.30}$$

where G is gravitational constant and M is the mass of the black hole. Moreover, physically meaningful solutions have approximate energy (in the sense of (2.20)) still equal to E. As a consequence, we are led to the study of the problem

$$\begin{cases} m\ddot{x} = -\left(mGM + \dfrac{4EGM}{c^2}\right)\dfrac{x}{|x|^3} - \dfrac{6G^2M^2}{c^2}\dfrac{x}{|x|^4} + \varepsilon\,\nabla U(x), \\ \dfrac{1}{2}m|\dot{x}|^2 - \left(mGM + \dfrac{4EGM}{c^2}\right)\dfrac{1}{|x|} - \dfrac{3G^2M^2}{c^2}\dfrac{1}{|x|^2} - \varepsilon\,U(x) = E. \end{cases} \tag{2.31}$$

Periodic solutions of this problem can be directly obtained from an application of Theorem 2.3, provided the parameter $\kappa(E, c)$ given in (2.30) is positive, i.e., when $E > -mc^2/4$. Indeed, fixed any energy E in this range, Theorem 2.3 yields solutions of any energy $E' < 0$ and so, in particular, of energy E. We then have the following result.

Corollary 2.1 *Let $E \in (-mc^2/4, 0)$ and $N \in \mathbb{N}$. Then, for every $\varsigma > 0$ there exists $\varepsilon^* = \varepsilon^*(N, \varsigma) > 0$ such that for every $\varepsilon \in \mathbb{R}$, with $|\varepsilon| \leq \varepsilon^*$, problem (2.31) has at least $2N$ periodic solutions $x_1^i, \ldots, x_N^i$, $i = 1, 2$.*

Moreover, for every $j = 1, \ldots, N$ and $i = 1, 2$, the solution x_j^i has minimal period $T_{i,j}$ satisfying

$$\left| T_{i,j} - \frac{\pi\kappa(E, c)\sqrt{m}}{\sqrt{2}(-E)^{3/2}} \right| < \varsigma$$

and winding number in its minimal period

$$\left\lfloor \frac{\sqrt{-2E}}{\kappa(E, c)} \sqrt{2\lambda(E, c) + \frac{\kappa(E, c)^2}{-2E}} \right\rfloor + j.$$

2.3.2 *The Kepler Problem with Relativistic Differential Operator*

In this section we deal with the problem

$$\frac{\mathrm{d}}{\mathrm{d}t}\left(\frac{m\dot{x}}{\sqrt{1 - |\dot{x}|^2/c^2}}\right) = -\alpha\,\frac{x}{|x|^3} + \varepsilon\,\nabla U(x), \qquad x \in \mathbb{R}^2 \setminus \{0\}, \tag{2.32}$$

where $m > 0$, $\alpha > 0$, $U \in C^\infty(\mathbb{R}^2, \mathbb{R})$, and $\varepsilon \in \mathbb{R}$. Notice that in the physical interpretation of (2.32) the constant α equals to mGM, where M is again the mass of the attracting body.

We first recall that for solutions of (2.32) an energy conservation law is fulfilled: Indeed, every solution x of (2.32) satisfies

$$E_\varepsilon(x, \dot{x}) = mc^2 \left(\frac{1}{\sqrt{1 - |\dot{x}|^2/c^2}} - 1 \right) - \frac{\alpha}{|x|} - \varepsilon\, U(x) = H, \tag{2.33}$$

for some $H \in \mathbb{R}$.

Now, we are in a position to state and prove our result for (2.32).

Theorem 2.4 *Let $H^* \in (-mc^2, 0)$ and $N \in \mathbb{N}$. Then, for every $\varsigma > 0$ there exists $\varepsilon^* = \varepsilon^*(N, \varsigma) > 0$ such that for every $\varepsilon \in \mathbb{R}$, with $|\varepsilon| \le \varepsilon^*$, Eq. (2.32) has at least $2N$ periodic solutions $x_1^i, \dots, x_N^i$, $i = 1, 2$, of energy H^*.*

Moreover, for every $j = 1, \dots, N$ and $i = 1, 2$, the solution x_j^i has minimal period $T_{i,j}$ satisfying

$$\left| T_{i,j} - \frac{2\pi m^2 c^3}{(-2mc^2 H^* - (H^*)^2)^{3/2}} \right| < \varsigma$$

and winding number $\lfloor mc^2/(H^* + mc^2) \rfloor + j$ in its minimal period.

Proof The proof is an application of Theorem 2.2, and it is based on the discussion and results in [5]. Please notice that in [5] the energy was defined by

$$\frac{mc^2}{\sqrt{1 - |\dot{x}|^2/c^2}} - \frac{\alpha}{|x|} - \varepsilon\, U(x),$$

thus implying that it differs from the energy E_ϵ defined in (2.33) by the additive factor mc^2. As a consequence, the formulas in [5] involving the energy have to be modified taking into account this additive factor.

For the proof, we adopt the same strategy of the proof of Theorem 2.3, and we outline the various steps, by pointing out the possible differences.

We first recall (cf. [5, Section 2]) that (2.32) can be written as a Hamiltonian system, with respect to the variables (x, p), where

$$p = \frac{m\dot{x}}{\sqrt{1 - \dfrac{|\dot{x}|^2}{c^2}}}.$$

The Hamiltonian is given by

$$\mathcal{H}_\varepsilon(x,p) = mc^2\left(\sqrt{1+\frac{|p|^2}{m^2c^2}}-1\right) - \frac{\alpha}{|x|} - \varepsilon\, U(x),$$

and the angular momentum $\mathcal{L}_0$ is again defined by (2.21).

Action-angles variables are defined as in the previous case, first introducing the change of variables

$$\Psi\colon \Omega \to (\mathbb{R}^2\setminus\{0\})\times\mathbb{R}^2, \qquad \Psi(r,\vartheta,l,\Phi) = \left(re^{i\vartheta}, le^{i\vartheta} + \frac{\Phi}{r} ie^{i\vartheta}\right),$$

being $\Omega = (0,+\infty)\times\mathbb{T}^1\times\mathbb{R}^2$. Then, defining

$$\Lambda = \left\{(H,L)\in\mathbb{R}^2\colon\ -mc^2 < H < 0,\ \frac{\alpha^2}{c^2} < L^2 < \frac{\alpha^2m^2c^2}{-2mc^2H - H^2}\right\},$$

for every $(H,L)\in\Lambda$ the orbit in the (r,l)-plane is closed. We point out that in this case the radial period map and the apsidal angle are given by

$$T(H,L) = \frac{2\pi m^2c^3}{(-2mc^2H - H^2)^{3/2}} \tag{2.34}$$

and

$$\Theta(H,L) = \frac{2\pi}{\sqrt{1-\dfrac{\alpha^2}{c^2L^2}}}, \tag{2.35}$$

for every $(H,L)\in\Lambda$, respectively.

Proceeding as in the case of the Levi-Civita potential, we introduce action-angle variables $(I_1, I_2, \varphi_1, \varphi_2)$, and in [5, Section 2.4] it is proved that in the new variables the Hamiltonian corresponding to the unperturbed case $\varepsilon = 0$ is given by

$$\mathcal{K}_0(I_1,I_2) = mc^2\frac{I_1 - I_2 + \frac{1}{c}\sqrt{c^2I_2^2-\alpha^2}}{\sqrt{(I_1-I_2)^2 + I_2^2 + \frac{2}{c}(I_1-I_2)\sqrt{c^2I_2^2-\alpha^2}}} - mc^2. \tag{2.36}$$

Now, let us fix $H^*\in(-mc^2,0)$, $N\in\mathbb{N}$, $j\in\{1,\ldots N\}$ and let us set $k_j^* = \lfloor mc^2/(H^*+mc^2)\rfloor + j$. Then, there exists a unique $L_j^* > 0$ such that

$$\sqrt{1-\frac{\alpha^2}{c^2(L_j^*)^2}} = \frac{1}{k_j^*}. \tag{2.37}$$

A simple computation shows that $(H^*, L_j^*) \in \Lambda$. Moreover, recalling (2.35), condition (2.37) implies that

$$\Theta(H^*, L^*) = 2\pi k_j^*,$$

and then the solution of the unperturbed problem corresponding to the pair (H^*, L_j^*) is periodic of period $2\pi m^2 c^3/(-2mc^2 H^* - (H^*)^2)^{\frac{3}{2}}$ (recall (2.34)), and it has winding number k_j^* in its period.

Let us now denote by I_j^* the pair of actions associated to (H^*, L_j^*) via the map Π defined in (2.25). The validity of (2.18), with $\mathcal{K}_0$ as in (2.36), follows from the fact that $\nabla^2 \mathcal{K}_0(I_j^*)$ is negative defined (cf. [5, Section 3.2]).

We then conclude as in the proof of Theorem 2.3. □

Remark 2.5 Let us notice that, analogously as in [5], one could also consider values (H^*, L^*) giving rise to periodic solutions of the unperturbed problem with minimal period $nT(H^*, L^*)$ with $n > 1$ (this happens if and only if $\Theta(H^*, L^*) \in 2\pi(\mathbb{Q} \setminus \mathbb{Z})$). The nondegeneracy condition can be verified in a similar manner, and thus a more general version of Theorem 2.4 could be proved. The same remark is also valid for the Levi-Civita problem discussed in Sect. 2.3.1. ◁

Acknowledgments Work written under the auspices of the Gruppo Nazionale per l'Analisi Matematica, la Probabilità e le loro Applicazioni (GNAMPA) of the Istituto Nazionale di Alta Matematica (INdAM).

References

1. A. Ambrosetti, U. Bessi, *Multiple periodic trajectories in a relativistic gravitational field*, Variational methods (Paris, 1988). Progr. Nonlinear Differential Equations Appl., 4, (Birkhäuser Boston, Boston, MA, 1990), pp. 373–381.
2. C. Andersen, H. von Baeyer, On classical scalar field theories and the relativistic Kepler problem. Ann. Phys. **62**, 120–134 (1971)
3. V. Arnol'd, *Mathematical methods of classical mechanics*. Graduate Texts in Mathematics (vol. 60) (Springer, New York, 1989)
4. G. Benettin, F. Fassò, *Introduzione alla teoria delle perturbazioni per sistemi Hamiltoniani* (2001–2002). Lecture notes, https://www.math.unipd.it/~benettin/postscript-pdf/hamilt.pdf
5. A. Boscaggin, W. Dambrosio, G. Feltrin, Periodic solutions to a perturbed relativistic Kepler problem. SIAM J. Math. Anal., **53**, 5813–5834 (2021)
6. A. Boscaggin, W. Dambrosio, G. Feltrin, *Periodic perturbations of central force problems and an application to a restricted 3-body problem*. J. Math. Pures Appl., **186**, 31–73 (2024)
7. A. Boscaggin, W. Dambrosio, D. Papini, *Periodic solutions to relativistic Kepler problems: a variational approach*. Ann. Sc. Norm. Super. Pisa Cl. Sci., https://doi.org/10.2422/2036-2145.202205_016
8. A. Boscaggin, W. Dambrosio, D. Papini, *Infinitely many periodic solutions to a Lorentz force equation with singular electromagnetic potential* (2023). arXiv:2302.06189
9. T. Boyer, Unfamiliar trajectories for a relativistic particle in a Kepler or Coulomb potential. Am. J. Phys. **72**, 992–997 (2004)

10. H. Broer, G. Huitema, A proof of the isoenergetic KAM-theorem from the "ordinary" one. J. Differ. Equ. **90** , 52–60 (1991)
11. B. Cordani, *The Kepler problem. Group theoretical aspects, regularization and quantization, with application to the study of perturbations*. Progress in Mathematical Physics (vol. 29) (Birkhäuser Verlag, Basel, 2003)
12. A. Fonda, A. Gallo, Radial periodic perturbations of the Kepler problem, Celestial Mechanics & Dynamical Astronomy. Int. J. Space Dynam. **129**, 257–268 (2017)
13. H. Goldstein, C. Poole, J. Safko, *Classical Mechanics* (Addison Wesley, San Francisco, 2002)
14. L. Jia, Approximate Kepler's elliptic orbits with the relativistic effects. Int. J. Astron. Astrophys. **3**, 29–33 (2013)
15. E. Lacomba, J. Llibre, A. Nunes, *Invariant tori and cylinders for a class of perturbed Hamiltonian systems*. The geometry of Hamiltonian systems (Berkeley, CA, 1989). Math. Sci. Res. Inst. Publ. (vol. 22) (Springer, New York, 1991), pp. 373–385
16. T. Lemmon, A. Mondragon, *Kepler's orbits and Special Relativity in Introductory Classical Mechanics* (2010). arXiv:1012.5438
17. T. Levi-Civita, *Fondamenti di Meccanica Relativistica* (Zanichelli, Bologna, 1928), p. 224
18. J. Mawhin, *Resonance problems for some non-autonomous ordinary differential equations*. Stability and bifurcation theory for non-autonomous differential equations. Lecture Notes in Math. (vol. 2065) (Springer, Heidelberg 2013), pp. 103–184
19. G. Muñoz, I. Pavic, A Hamilton-like vector for the special-relativistic Coulomb problem. Eur. J. Phys. **27**, 1007–1018 (2006)
20. A. Nunes, J. Casasayas, J. Llibre, *A perturbation of the relativistic Kepler problem*. Predictability, stability, and chaos in N-body dynamical systems (Cortina d'Ampezzo, 1990). NATO Adv. Sci. Inst. Ser. B: Phys. (vol. 272) (Plenum, New York, 1991), pp. 547–554
21. H. Pollard,*Celestial Mechanics*. Carus Mathematical Monographs (vol. 18) (Mathematical Association of America, Washington, DC, 1976)
22. P. Torres, A. Ureña, M. Zamora, Periodic and quasi-periodic motions of a relativistic particle under a central force field. Bull. Lond. Math. Soc. **45**, 140–152 (2013)
23. A. Weinstein, Normal modes for nonlinear Hamiltonian systems. Inventiones Mathematicae **20**, 47–57 (1973)
24. A. Weinstein, Symplectic V-manifolds, periodic orbits of Hamiltonian systems, and the volume of certain Riemannian manifolds. Commun. Pure Appl. Math. **30**, 265–271 (1977)
25. A. Weinstein, Bifurcations and Hamilton's principle. Mathematische Zeitschrift **159**, 235–248 (1978)
26. M. Zamora, New periodic and quasi-periodic motions of a relativistic particle under a planar central force field with applications to scalar boundary periodic problems. Electron. J. Qualitat. Theory Differ. Equ. **31**, 1–16 (2013)

Chapter 3
A Survey on Some Existence Results for the Relativistic Pendulum Equation

J. Ángel Cid

Mathematics Subject Classification 34C25, 83A05, 49J40, 47H11

3.1 Introduction

A very well-known and important model that can be described by a second-order differential equation is the simple pendulum: A bob of mass $m > 0$ is joined to a fixed pivot O through a rigid string with negligible mass and length $l > 0$. We assume that the only forces acting on the bob are:

1. A constant gravitational force of magnitude g in the direction of a unitary vector $\mathbf{e} \in \mathbb{R}^3$.
2. The tension T of the string that maintains always the bob at the same distance l of the pivot.
3. A friction term proportional to the velocity of the bob.

Partially supported by the Agencia Estatal de Investigación (Ministerio de Ciencia e Innovación), Spain, and European Regional Development Fund, Project PID2021-122442NB–100.

J. Ángel Cid (✉)
CITMAga, Santiago de Compostela, Spain

Departamento de Matemáticas, Universidade de Vigo, Campus de Ourense, Spain
e-mail: angelcid@uvigo.es

P. Amster, P. Benevieri (eds.), *Topological Methods for Delay and Ordinary Differential Equations*, Advances in Mechanics and Mathematics 51,
https://doi.org/10.1007/978-3-031-61337-1_3

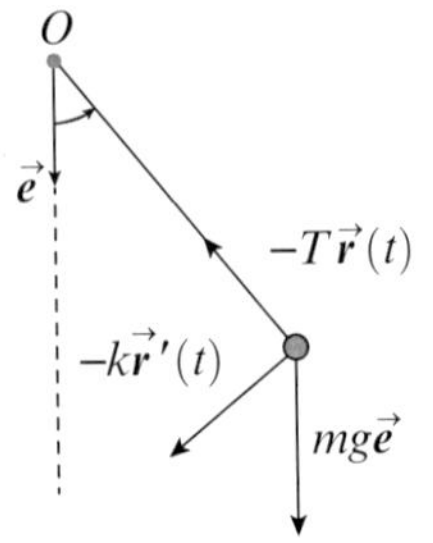

Then, Newton's second law tells us that the position of the pendulum $\mathbf{r}(t)$, measured from the pivot O, satisfies the differential equation

$$\mathbf{F} = \frac{d}{dt}\left(m \cdot \mathbf{r}'(t)\right), \tag{3.1}$$

where

$$\mathbf{F} = mg\mathbf{e} - T\mathbf{r}(t) - km\mathbf{r}'(t). \tag{3.2}$$

In the classical Newtonian setting the mass m is a constant, so (3.1) and (3.2) lead to

$$\mathbf{r}'' = g\mathbf{e} - \frac{T}{m}\mathbf{r}(t) - k\mathbf{r}'(t). \tag{3.3}$$

Now, if the initial velocity $\mathbf{r}'(0)$ lies in the same plane generated by $\mathbf{e}$ and the initial position of the bob $\mathbf{r}(0)$, then the motion of the pendulum is plane: Indeed, let $\mathbf{u}$ a normal vector to both $\mathbf{e}$, $\mathbf{r}(0)$, and $\mathbf{r}'(0)$ and define

$$h(t) = \langle \mathbf{r}(t), \mathbf{u} \rangle,$$

where $\langle \cdot, \cdot \rangle$ is the usual scalar product in $\mathbb{R}^3$. By assumption $h(0) = 0$, $h'(0) = 0$, and

$$h''(t) = \langle \mathbf{r}''(t), \mathbf{u} \rangle = \langle g\mathbf{e} - \frac{T}{m}\mathbf{r}(t) - k\mathbf{r}'(t), \mathbf{u} \rangle = -\frac{T}{m}h(t) - kh'(t).$$

That is, $h(t)$ is the solution of an homogeneous linear differential equation through the origin, thus $h \equiv 0$ as we wanted to show.

Let us consider $\{\mathbf{e_1}, \mathbf{e_2}\}$ an orthonormal basis of $\mathbf{u}^{\perp}$ with $\mathbf{e_1} = \mathbf{e}$. Then we can decompose $\mathbf{r}(t) = r_1(t)\mathbf{e_1} + r_2(t)\mathbf{e_2}$ where

$$\begin{cases} r_1''(t) = g - kr_1'(t) - \frac{T}{m}r_1(t), \\ r_2''(t) = -kr_2'(t) - \frac{T}{m}r_2(t). \end{cases}$$

Since $\|\mathbf{r}(t)\| = l > 0$ for all t, we can write the components $r_1(t)$ and $r_2(t)$ in polar coordinates as

$$\begin{cases} r_1(t) = l\cos(\theta(t)), \\ r_2(t) = l\sin(\theta(t)), \end{cases}$$

where

$$\theta' = \frac{1}{l^2}(r_1 r_2' - r_1' r_2).$$

By deriving again the above expression, we get that $\theta(t)$ should satisfy the second-order scalar differential equation

$$\theta'' + k\theta' + \frac{g}{l}\sin(\theta) = 0. \tag{3.4}$$

Reciprocally, if $\theta(t)$ satisfies (3.4), there exists a unique expression for T, namely

$$T = \frac{mg\cos(\theta)}{l} + m\theta'^2$$

such that $\mathbf{r}(t) = l\cos(\theta(t))\mathbf{e_1} + l\sin(\theta(t))\mathbf{e_2}$ is indeed a solution of (3.3). Notice that given two solutions θ_1, θ_2 of (3.4), they provide the same solution $\mathbf{r}(t)$ of (3.3) if and only if $\theta_2 \equiv \theta_1 + 2m\pi$, where $m \in \mathbb{Z}$. Therefore we will say that θ_1 and θ_2 are *geometrically different solutions* of (3.4) when they do not differ by an integer multiple of 2π.

In this chapter we will write (3.4) as

$$x'' + kx' + a\sin(x) = 0, \tag{3.5}$$

where $k \geq 0$ and $a > 0$, and we will call (3.5) the *classical free pendulum equation* that can be conservative, $k = 0$, or to include a damping term, $k > 0$. Of course both cases are qualitatively well understood, and the conservative one can even be analytically solved by means of elliptic functions. However, when an external force is added, we get the much more difficult *classical forced pendulum* equation

$$x'' + kx' + a\sin(x) = f(t) \tag{3.6}$$

that has been a true landmark for the nonlinear analysis in the last century as it was emphasized in several nice surveys by J. Mawhin [14–16].

Much more recently is the study of a relativistic counterpart of (3.6) which is obtained by assuming in the right-hand side of (3.1) a basic principle in the theory of Einstein's special relativity: The mass m of a moving body along the trajectory $\mathbf{r}(t)$ is no longer constant, but it depends on the modulus of its velocity, namely

$$m = \frac{m_0}{\sqrt{1 - \frac{\|\mathbf{r}'(t)\|^2}{\mathfrak{c}^2}}},$$

where m_0 is the mass of the body at rest and $\mathfrak{c}$ is the speed of the light in the vacuum. Of course in the relativistic setting one has the restriction $\|\mathbf{r}'(t)\| < \mathfrak{c}$.

The *relativistic forced pendulum equation*, corresponding to the formulation (3.6) in the Newtonian mechanics, is thus given by

$$\left(\frac{x'}{\sqrt{1 - \frac{x'^2}{c^2}}}\right)' + kx' + a\sin(x) = f(t), \tag{3.7}$$

where $c, a > 0$ and $k \geq 0$.

Note that from the mathematical point of view both equations (3.6) and (3.7) fit into the framework of the so-called ϕ-Laplacian equations

$$\phi(x')' + kx' + a\sin(x) = f(t), \tag{3.8}$$

where $\phi : (-A, A) \to (-B, B)$ is an homeomorphism and $0 < A, B \leq +\infty$, see for instance [3, 4, 9, 10, 22].

The goal of this survey is to present some recent results on the existence of T-periodic solutions (from now on $T > 0$ will represent the fixed value of the period) for (3.7), emphasizing the dynamical similarities and differences with respect to the classical setting (3.6). Be aware that most of the results given in the literature, for instance those in [2, 3, 6, 9], are in fact more general than the versions we present here: We have particularized them to equation (3.7) and adapted the proofs to this context.

Through the text we will use the following notation:

$$C := \{x\colon [0, T] \to \mathbb{R}\colon\ x \text{ is continuous in } [0, T]\},$$

$$C^1 := \{x\colon [0, T] \to \mathbb{R}\colon\ x \text{ is continuously differentiable in } [0, T]\},$$

$$C_T := \{x \in C\colon x(0) = x(T)\},$$

$$C_T^1 := \{x \in C^1\colon x(0) = x(T), \quad x'(0) = x'(T)\},$$

$$\bar{f} := \frac{1}{T}\int_0^T f(s)ds \quad \text{and} \quad \tilde{f} = f - \bar{f} \quad \text{for any } f \in L^1[0, T],$$

$$\tilde{L}^1 := \{f \in L^1[0, T]\colon \bar{f} = 0\},\ \tilde{C} := \{f \in C\colon \bar{f} = 0\} \text{ and so on.}$$

3.2 The Conservative Case $k = 0$

3.2.1 *The Classical Framework*

The existence of a solution for the periodic boundary value problem associated to the conservative classical forced pendulum

$$x'' + a\sin(x) = f(t), \quad x(0) = x(T), \quad x'(0) = x'(T), \tag{3.9}$$

where $a, T > 0$ and $f \in C$, was firstly settled by Hamel in 1922, [13]. Notice that integrating the equation and taken into account the periodicity, we get

$$\frac{1}{T}\int_0^T f(s)ds \in [-a, a],$$

so $\bar{f} = 0$ is a necessary condition in order problem (3.9) has a solution for all $a > 0$.

Hamel used the direct method of the calculus of variations to prove that $\bar{f} = 0$ is also a sufficient condition for the solvability of (3.9) by showing that the action functional $\mathcal{J}\colon C_T^1 \to \mathbb{R}$ defined by

$$x \in C_T^1 \to \mathcal{J}x := \int_0^T \frac{(x'(t))^2}{2} + a\cos(x(t)) + x(t)f(t)dt$$

has a minimum over C_T^1.

Actually he proved his result for functions of the form $f(t) = b\cos(\omega t)$, but his method worked as well for any $f \in \tilde{C}$. This result was rediscovered, almost sixty years later, independently by Castro et al. [8, 11, 23].

3.2.2 *The Relativistic Framework*

In 2010 Brézis and Mawhin, [6, 17], carried over Hamel's result to the framework of the relativistic forced pendulum

$$\left(\frac{x'}{\sqrt{1-\frac{x'^2}{c^2}}}\right)' + a\sin(x) = f(t), \quad x(0) = x(T), \quad x'(0) = x'(T), \tag{3.10}$$

where $f \in \tilde{L}^1$ and a solution means that $x \in C_T^1$, $\|x'\|_\infty < c$, $\dfrac{x'}{\sqrt{1-\frac{x'^2}{c^2}}}$ is absolutely continuous on $[0, T]$ and satisfies (3.10).

The action functional is now defined on a subset of the periodic Lipschitz functions

$$Lip_T := \{x\colon [0,T] \to \mathbb{R}:\ x \text{ is Lipschitzian and } x(0) = x(T)\}.$$

Notice that x is Lipschitzian if and only if x is absolutely continuous and $x' \in L^\infty$. Moreover, it is easy to show that

$$\|x'\|_\infty = \sup_{t,s\in[0,T],t\neq s} \frac{x(t)-x(s)}{t-s},$$

and Lip_T is a Banach space with the norm

$$\|x\|_{Lip} := \|x\|_\infty + \|x'\|_\infty.$$

Let us consider now the closed convex set

$$K := \{x \in Lip_T\colon \|x'\|_\infty \le c\},$$

and the action functional $\mathcal{J}_r\colon K \to \mathbb{R}$ defined by

$$\mathcal{J}_r(x) = \int_0^T \left(-c^2\sqrt{1-\frac{x'(t)^2}{c^2}} + a\cos(x(t)) + f(t)x(t)\right) dt, \tag{3.11}$$

and which is C^1 on the interior of K. So, the objective is to find a minimum, x, of $\mathcal{J}_r$ over K that moreover satisfies $\|x'\|_\infty < c$.

The following technical lemma is useful in order to prove the lower semicontinuity of $\mathcal{J}_r$.

Lemma 3.1 *If $\{x_n\} \subset K$ converges uniformly on $C[0,T]$ to some $x \in K$, then*

$$\liminf_{n\to\infty} \int_0^T -c^2\sqrt{1-\frac{x_n'(t)^2}{c^2}}dt \ge \int_0^T -c^2\sqrt{1-\frac{x'(t)^2}{c^2}}dt.$$

Proof Notice that the function

$$z \in [-c,c] \to \Phi(z) = -c^2\sqrt{1-\frac{z^2}{c^2}}$$

satisfies

$$\Phi'(z) = \frac{z}{\sqrt{1-\frac{z^2}{c^2}}} \quad \text{and} \quad \Phi''(z) = \frac{1}{\left(1-\frac{z^2}{c^2}\right)^{3/2}} > 0 \quad \text{for all } z \in (-c,c),$$

so Φ is convex and for any $\lambda \in (0, 1)$ holds that

$$\int_0^T \Phi[x_n'(t)]dt \geq \int_0^T \Phi[\lambda x'(t)]dt + \int_0^T \Phi'[\lambda x'(t)][x_n'(t) - \lambda x'(t)]dt.$$

On the other hand, x_n' converges to x' in the weak-$*$ topology of L^∞, which implies

$$\lim_{n\to\infty} \int_0^T \Phi'[\lambda x'(t)]x_n'(t)dt = \int_0^T \Phi'[\lambda x'(t)]x'(t)dt.$$

Then, it follows that

$$\liminf_{n\to\infty} \int_0^T \Phi[x_n'(t)]dt \geq \int_0^T \Phi[\lambda x'(t)]dt + (1-\lambda) \int_0^T \Phi'[\lambda x'(t)]x'(t)dt.$$

Finally, since

$$\int_0^T \Phi'[\lambda x'(t)]x'(t)tdt = \int_0^T \frac{\lambda x'(t)^2}{\sqrt{1 - \frac{\lambda^2 x'(t)^2}{c^2}}}dt \geq 0,$$

the inequality

$$\liminf_{n\to\infty} \int_0^T \Phi[x_n'(t)]dt \geq \int_0^T \Phi[x'(t)]dt$$

holds by taking the limit when $\lambda \to 1^-$ and using the Lebesgue dominated convergence theorem. □

Now it is easy to show that $\mathcal{J}_r$ attains a minimum on K for any L^1-forcing f with mean value zero.

Theorem 3.1 *If $f \in \tilde{L}^1$, then $\mathcal{J}_r$ has a minimum over K.*

Proof Because of the 2π-periodicity of $\mathcal{J}_r$, i.e., $\mathcal{J}_r(x + 2\pi) = \mathcal{J}_r(x)$, one can minimize, without loss of generality, $\mathcal{J}_r$ in the convex set

$$\hat{K} = \{x \in K : \bar{x} \in [0, 2\pi]\}.$$

Notice that if $x \in \hat{K}$, from the Sobolev inequality $\|\tilde{x}\|_\infty^2 \leq \frac{T}{12}\int_0^T |x'(t)|^2 dt$, we have

$$\|x\|_\infty \leq |\bar{x}| + \|\tilde{x}\|_\infty \leq 2\pi + \frac{cT}{2\sqrt{3}},$$

so $\hat{K}$ is a bounded and equicontinuous set in C_T.

Now, let $\{x_n\}$ be a minimizing sequence for $\mathcal{J}_r$ in $\hat{K}$. By the Arzelà–Ascoli's theorem one can assume, passing to a subsequence if necessary, that $\{x_n\}$ converges uniformly to a function $x \in C_T$. Then it is easy to show that actually $x \in \hat{K}$ and hence by the previous lemma

$$\inf_{\hat{K}} \mathcal{J}_r = \lim_{n\to\infty} \mathcal{J}_r(x_n) \geq \mathcal{J}_r(x).$$

□

As pointed out in [6], the minimizer x of $\mathcal{J}_r$ in K should satisfy also $\|x'\| < c$ in order to be a solution of (3.10). With this goal in mind, they prove firstly that any minimizer x satisfies for all $y \in Lip_T$ the variational inequality

$$\int_0^T \left(-c^2\sqrt{1-\frac{y'(t)^2}{c^2}} + c^2\sqrt{1-\frac{x'(t)^2}{c^2}} + (f(t) - a\sin(x(t)))(y(t)-x(t)) \right) dt \geq 0,$$

and for any $h \in L^1$ the auxiliary problem

$$\left(\frac{x'}{\sqrt{1-\frac{x'^2}{c^2}}} \right)' - x(t) = h(t), \quad x(0) = x(T), \quad x'(0) = x'(T),$$

has a unique solution $\hat{x}_h$ and $\|\hat{x}_h'\|_\infty < c$. Furthermore, for any $y \in K$ holds that

$$\int_0^T \left(-c^2\sqrt{1-\frac{y'(t)^2}{c^2}} + c^2\sqrt{1-\frac{\hat{x}_h'(t)^2}{c^2}} + (\hat{x}_h(t) + h(t))(y(t)-\hat{x}_h(t)) \right) dt \geq 0.$$

As a consequence of the preceding statements, Brézis and Mawhin were able to reproduce Hamel's result in the context of the relativistic forced pendulum without friction.

Theorem 3.2 ([6]) *For any $f \in \tilde{L}^1$ and $a > 0$ the T-periodic boundary value problem* (3.10) *has a solution that minimizes $\mathcal{J}_r$ over K.*

Proof Let x a minimizer of $\mathcal{J}_r$ over K that exists by Theorem 3.1, and consider

$$h(t) = f(t) - a\sin(x(t)) - x(t) \in L^1.$$

From the variational inequalities satisfied by x and $\hat{x}_h$, applied respectively to $y = \hat{x}_h$ and $y = x$, one gets

$$\int_0^T \left(x(t) - \hat{x}_h(t)\right)^2 dt \leq 0,$$

so $x = \hat{x}_h$, and then $\|x'\|_\infty = \|\hat{x}_h'\|_\infty < c$ and x is a solution of (3.10). □

We point out that a second geometrically different T-periodic solution for the classical forced pendulum without friction was obtained by means of the mountain pass lemma by Mawhin and Willem, [18], more than sixty years after Hamel. However, it took only two years after [6] to show the same result in the relativistic framework by Bereanu and Torres [5].

So, in the conservative case, $k = 0$, the periodic dynamics of the relativistic pendulum, although technically more demanding, mimics that of the classical one. However, we will see in the next section that this is no longer true for the nonconservative pendulum.

3.3 The Nonconservative Case $k > 0$

3.3.1 A Pioneering Result

The results in the previous section could give the impression that problem (3.6) always has a T-periodic solution for any $a, k > 0$ and $f \in \tilde{C}_T$. However it was shown by Ortega that this is not the case, [19], since for each $a, k, T > 0$ there exists $f \in \tilde{C}_T$ such that problem (3.6) has no solution, see also [1, 20].

In a seminal paper, [21], Torres pointed out a striking difference between problems (3.6) and (3.7) by showing that for small periods (3.7) always has a T-periodic solution for any $a, k > 0$ and $f \in \tilde{C}_T$. Actually, it was precisely [21] that started the recent interest in the relativistic pendulum equation.

By making the change of variables $x = \arcsin y$ the equation (3.7) is equivalent to

$$\phi\left(\frac{y'}{\sqrt{1-y^2}}\right)' + k\frac{y'}{\sqrt{1-y^2}} + ay = f(t), \tag{3.12}$$

where $\phi : (-c, c) \to \mathbb{R}$ is given by

$$\phi(z) = \frac{z}{\sqrt{1-\frac{z^2}{c^2}}}. \tag{3.13}$$

By integrating twice one see that any solution y of (3.12) satisfies also the integral equation

$$y(t) = \int_0^t \sqrt{1-y^2}\phi^{-1}\left(F[y](s) + C\right) ds + D,$$

where

$$F[y](s) = \int_0^t (f(s) - ay(s))ds - k\arcsin(y)$$

and $C, D \in \mathbb{R}$ are arbitrary and will be fixed in the following result.

Proposition 3.1 ([21]) *For any* $y \in K := \{y \in \tilde{C}_T : \|y\|_\infty \leq 1\}$, *there exists a unique choice of* C_y *and* D_y *such that*

$$\mathcal{T}[y](t) \equiv \int_0^t \sqrt{1-y^2}\phi^{-1}\left(F[y](s)+C_y\right)ds - D_y \in \tilde{C}_T$$

and holds that

$$D_y = -\overline{\int_0^{\cdot} \sqrt{1-y^2}\phi^{-1}\left(F[y](s)+C_y\right)ds}.$$

Furthermore, the operator $\mathcal{T} : K \to \tilde{C}_T$ *is completely continuous, and its fixed points in the interior of* K *are* T*-periodic solutions of* (3.12).

Now, it is immediate to prove the main result in [21].

Theorem 3.3 *If* $cT \leq \dfrac{1}{2}$ *then for any values* $a, k > 0$ *and* $f \in \tilde{C}_T$, *the problem* (3.7) *has at least a* T*-periodic solution.*

Proof Since $\|\mathcal{T}[y]\|_\infty < 2cT \leq 1$ for all $y \in K$, we have that the closed, convex, and nonempty set K is invariant for the completely continuous operator $\mathcal{T}$, so Schauder's fixed point theorem implies the existence of a fixed point $y \in K$ of $\mathcal{T}$ with $\|y\|_\infty < 1$. Then y is a T-periodic solution of (3.12), and we can define $x = \arcsin(y)$ that is a T-periodic solution of (3.7). □

3.3.2 The Search for Better Estimates

Probably, the most fundamental open question with respect to problem (3.7) is to determine if it always has a T-periodic solution independently on the size of $T > 0$. Until this question can be settled, a fruitful line of research has been to improve the quantitative estimate in the assumption of Theorem 3.3, starting by Bereanu, Jebelean and Mawhin who in [3] obtained by means of degree theory the following improvement:

$$cT \leq \sqrt{3}\pi.$$

Let us show their approach for the periodic boundary value problem

$$\phi\left(x'\right)' + kx' + a\sin(x) = f(t), \quad x(0) = x(T), \quad x'(0) = x'(T), \tag{3.14}$$

where $f \in \tilde{C}_T$ and ϕ is defined by (3.13).

Firstly we need the following simple technical result.

Lemma 3.2 ([4]) *For each $f \in C$ there exists a unique $\alpha = \alpha(f) \in Range(f)$ such that*

$$\int_0^T \phi^{-1}(f(t) - \alpha)\, dt = 0.$$

Moreover the function $f \in C \to \alpha \in \mathbb{R}$ is continuous.

Clearly, problem (3.14) can be rewritten as

$$\phi\left(x'\right)' = N_f(x), \quad x(0) = x(T), \quad x'(0) = x'(T), \tag{3.15}$$

where

$$N_f(x)(t) := f(t) - kx'(t) - a\sin(x(t)),$$

and it is easy to show that when $f \in \tilde{C}_T$ the solutions of (3.15) are the fixed points of the operator $\mathcal{M} : C_T^1 \to C_T^1$ defined as

$$\mathcal{M}x(t) := x(0) + \overline{N_f(x)} + \int_0^t \phi^{-1}\left[\int_0^s \left(N_f(x)(r) - \overline{N_f(x)}\right) dr - \alpha\right] ds.$$

Indeed, $\mathcal{M}$ is completely continuous in C_T^1, and notice that by Lemma 3.2

$$\begin{aligned}\mathcal{M}x(T) &= x(0) + \overline{N_f(x)} + \int_0^T \phi^{-1}\left[\int_0^s \left(N_f(x)(r) - \overline{N_f(x)}\right) dr - \alpha\right] ds \\ &= x(0) + \overline{N_f(x)} = \mathcal{M}x(0).\end{aligned}$$

On the other hand,

$$(\mathcal{M}x)'(t) = \phi^{-1}\left[\int_0^t \left(N_f(x)(r) - \overline{N_f(x)}\right) dr - \alpha\right] \quad \text{for } t \in [0, T],$$

so

$$\begin{aligned}(\mathcal{M}x)'(T) &= \phi^{-1}\left[\int_0^T \left(N_f(x)(r) - \overline{N_f(x)}\right) dr - \alpha\right] \\ &= \phi^{-1}\left[T\overline{N_f(x)} - T\overline{N_f(x)} - \alpha\right] = \phi^{-1}(-\alpha) = (\mathcal{M}x)'(0).\end{aligned}$$

Then $\mathcal{M}x \in C_T^1$, and clearly,

$$\phi((\mathcal{M}x)')'(t) = N_f(x)(t) - \overline{N_f(x)} \quad \text{for } t \in [0, T].$$

Now, take into account that if $x = \mathcal{M}x$, then in particular

$$x(0) = \mathcal{M}x(0) = x(0) + \overline{N_f(x)},$$

so $\overline{N_f(x)} = 0$ and hence x is a solution of (3.14).

Now, the key to show the existence of a fixed point for $\mathcal{M}$ in C_T^1 is the use of the Leray–Schauder degree combined with suitable bounds for the possible fixed points of the family of operators

$$\mathcal{M}_\lambda x(t) := x(0) + \overline{N_f(x)} + \int_0^t \phi^{-1}\left[\int_0^s \lambda\left(N_f(x)(r) - \overline{N_f(x)}\right) dr - \alpha\right] ds, \tag{3.16}$$

for $\lambda \in [0, 1]$ and which are the solutions of the modified problem

$$\phi\left(x'\right)' = \lambda(f(t) - kx'(t) - a\sin(x(t))),\ x(0) = x(T),\ x'(0) = x'(T). \tag{3.17}$$

Theorem 3.4 ([3]) *Assume that* $cT \leq \sqrt{3}\pi$. *Then for any values* $a, k > 0$ *and* $f \in \tilde{C}_T$ *the problem* (3.7) *has at least two geometrically different* T*-periodic solutions* x_1, x_2 *such that*

$$-\frac{\pi}{2} < \overline{x}_1 < \frac{\pi}{2} < \overline{x}_2 < \frac{3\pi}{2}.$$

Proof Let us consider the open and bounded set $\Omega_1 \subset C_T^1$ defined as

$$\Omega_1 := \left\{x \in C_T^1 : |\overline{x}| < \frac{\pi}{2}, \quad \|x'\|_\infty < c, \quad \|x - \overline{x}\|_\infty < \frac{cT}{2\sqrt{3}}\right\},$$

and the homotopic family $\mathcal{M}_\lambda : \overline{\Omega}_1 \to C_T^1$. Now the proof proceeds along the following standard steps:

Step 1. There is no solution of the equation $x = \mathcal{M}_\lambda x$ *for* $\lambda \in [0, 1]$ *and* $x \in \partial\Omega_1$.

Assume to the contrary that there exist $x \in \partial\Omega_1$ and $\lambda \in [0, 1]$ such that $x = \mathcal{M}_\lambda x$. Then $\|x'\|_\infty < c$, and from the Sobolev inequality

$$\|x - \overline{x}\|_\infty^2 \leq \frac{T}{12}\int_0^T |x'(t)|^2 dt,$$

we obtain

$$\|x - \overline{x}\|_\infty < \frac{cT}{2\sqrt{3}}.$$

So, as $x \in \partial\Omega_1$, the unique possibility is that $|\overline{x}| = \dfrac{\pi}{2}$. Let us suppose that $\overline{x} = \dfrac{\pi}{2}$ (the case $\overline{x} = -\dfrac{\pi}{2}$ is similar), then

$$\left\| x - \frac{\pi}{2} \right\|_\infty = \|x - \overline{x}\|_\infty < \frac{cT}{2\sqrt{3}} \leq \frac{\pi}{2},$$

and therefore $0 < x(t) < \pi$ for all $t \in [0, T]$. But now integrating the equation in (3.17) between 0 and T, we attain the contradiction

$$0 = \int_0^T \sin(x(s))ds > 0.$$

Step 2. $d_{LS}(Id - \mathcal{M}_0, \Omega) = 1$.

By *Step 1* the Leray–Schauder degree of $Id - \mathcal{M}_\lambda$ in Ω is well-defined for any $\lambda \in [0, 1]$. In case $\lambda = 0$ we have that the range of operator

$$\mathcal{M}_0(x) = x(0) + \overline{N_f(x)}$$

is contained in the subspace of constant functions that is isomorphic to $\mathbb{R}$, and then the reduction property of the Leray–Schauder degree, [12], allows us to compute

$$\begin{aligned} d_{LS}(Id - \mathcal{M}_0, \Omega) &= d_B\left((Id - \mathcal{M}_0)|_{\mathbb{R}}, \left(-\frac{\pi}{2}, \frac{\pi}{2}\right)\right) \\ &= \frac{\text{sign}\left(-\overline{N_f\left(\frac{\pi}{2}\right)}\right) - \text{sign}\left(-\overline{N_f\left(-\frac{\pi}{2}\right)}\right)}{2} \\ &= \frac{\text{sign}\left(a \sin\left(\frac{\pi}{2}\right)\right) - \text{sign}\left(a \sin\left(-\frac{\pi}{2}\right)\right)}{2} = 1, \end{aligned}$$

where d_B stands for the Brouwer degree.

Step 3. Problem (3.7) *has a* T*-periodic solution* x_1 *such that* $-\frac{\pi}{2} < \overline{x}_1 < \frac{\pi}{2}$. From the homotopy invariance of the Leray–Schauder degree and the previous step, we obtain

$$d_{LS}(Id - \mathcal{M}, \Omega) = d_{LS}(Id - \mathcal{M}_1, \Omega) = d_{LS}(Id - \mathcal{M}_0, \Omega) \neq 0.$$

Then, there exists a fixed point x_1 of $\mathcal{M}$ in Ω_1 that is a solution of (3.14) and consequently a solution of problem (3.7) with $-\frac{\pi}{2} < \overline{x}_1 < \frac{\pi}{2}$.

Now, the existence of a second T-periodic solution x_2 with $\frac{\pi}{2} < \overline{x}_2 < \frac{3\pi}{2}$ follows from a similar reasoning applied to the open bounded set

$$\Omega_2 := \left\{ x \in C_T^1 : \frac{\pi}{2} < \overline{x} < \frac{3\pi}{2}, \quad \|x'\|_\infty < c, \quad \|x - \overline{x}\|_\infty < \frac{cT}{2\sqrt{3}} \right\}.$$

□

By studying an equivalent system to (3.7) and using a continuation theorem, the author obtained in [9] the improvement

$$cT \leq 2\pi.$$

Indeed, by means of the change of variables $y = \phi(x') + kx$, to find a T-periodic solution of (3.7) is equivalent to solve the following periodic boundary value problem for a first-order system:

$$\begin{cases} x' = \phi^{-1}(y - kx), & x(0) = x(T), \\ y' = -a\sin(x) + f(t), & y(0) = y(T). \end{cases} \tag{3.18}$$

Let us consider the periodic BVP

$$z' = F(t, z), \quad z(0) = z(T), \tag{3.19}$$

assuming that

$$F(t, z) := f(t, z; 1),$$

where $f : [0, T] \times \mathbb{R}^n \times [0, 1] \to \mathbb{R}^n$ is a continuous function such that for $\lambda = 0$ the map f is autonomous, that is, there exists a continuous function $f_0 : \mathbb{R}^n \to \mathbb{R}^n$ such that

$$f_0(z) := f(t, z; 0),$$

for all $t \in [0, T]$ and all $z \in \mathbb{R}^n$.

Lemma 3.3 ([7]) *Let Ω be a bounded and open subset of $\mathbb{R}^n$ and suppose that the following conditions are satisfied:*

- *("Bound set" condition) For any $\lambda \in [0, 1)$ and any z solution of*

$$z' = f(t, z; \lambda), \quad z(0) = z(T), \tag{3.20}$$

  such that $z(t) \in \overline{\Omega}$ for all $t \in [0, T]$, it follows that $z(t) \in \Omega$ for all $t \in [0, T]$.
- *$d_B(f_0, \Omega, 0) \neq 0$, where d_B stands for the Brouwer degree in $\mathbb{R}^n$.*

Then, problem (3.19) has at least one solution $z(t)$ such that $z(t) \in \overline{\Omega}$ for all $t \in [0, T]$.

In order to apply Lemma 3.3 to problem (3.18) the following estimates are essential.

Lemma 3.4 ([9]) *Assume that $a, k > 0$, $f \in \tilde{C}_T$, and $\lambda \in [0, 1]$. If (x, y) is a solution of the modified problem*

$$\begin{cases} x' = \phi^{-1}(y - kx), & x(0) = x(T), \\ y' = -a\sin(x) + \lambda f(t), & y(0) = y(T) \end{cases} \tag{3.21}$$

such that

$$\|x - m\pi\|_\infty \le l, \quad \textit{for some } 0 < l < \pi \textit{ and } m \in \mathbb{Z},$$

then the following estimates hold:

(i) There exists $t_0 \in [0, T]$ *such that* $x(t_0) = m\pi$.
(ii) $\|y - km\pi\|_\infty \le kl + a\frac{T}{2} + \frac{\|f\|_{L^1}}{2}$.
(iii) $\|x - m\pi\|_\infty \le \max_{t\in[0,T]} x(t) - \min_{t\in[0,T]} x(t) \le \frac{T}{2}\phi^{-1}\left(2kl + a\frac{T}{2} + \frac{\|f\|_{L^1}}{2}\right)$.

Now, we present the main result in [9] with respect to the relativistic pendulum.

Theorem 3.5 ([9]) *Suppose that* $a, k, T > 0$, $f \in \tilde{C}_T$, *and*

$$\frac{T}{2}\phi^{-1}\left(2k\pi + a\frac{T}{2} + \frac{\|f\|_{L^1}}{2}\right) < \pi.$$

Then the problem (3.7) *has at least two geometrically different* T*-periodic solutions,* x_1 *and* x_2, *such that:*

$$-\pi < x_1(t) < \pi \textit{ for all } t \in \mathbb{R} \textit{ and } x_1(t_1) = 0 \textit{ for some } t_1 \in [0, T], \tag{3.22}$$

$$0 < x_2(t) < 2\pi \textit{ for all } t \in \mathbb{R} \textit{ and } x_2(t_2) = \pi \textit{ for some } t_2 \in [0, T]. \tag{3.23}$$

Proof By the assumption there exists $l > 0$ such that

$$\frac{T}{2}\phi^{-1}\left(2k\pi + a\frac{T}{2} + \frac{\|f\|_{L^1}}{2}\right) < l < \pi.$$

Let us consider the open bounded set in $\mathbb{R}^2$

$$\Omega_1 = \left\{(x, y) \in \mathbb{R}^2 : |x| < l,\ |y| < k\pi + a\frac{T}{2} + \frac{\|f\|_{L^1}}{2}\right\}.$$

Now, our strategy to find a T-periodic solution x_1 satisfying (3.22) is to apply Lemma 3.3 to the homotopic system (3.21) and the set Ω_1. So it is enough to verify the following two claims.

Claim 1. For any $\lambda \in [0, 1]$ *and any* (x, y) *solution of* (3.21) *such that* $(x(t), y(t)) \in \overline{\Omega}_1$ *for all* $t \in [0, T]$, *it follows that* $(x(t), y(t)) \in \Omega_1$ *for all* $t \in [0, T]$.

Since $(x(t), y(t)) \in \overline{\Omega}_1$ for all $t \in [0, T]$, then (x, y) satisfies the assumptions in Lemma 3.4 with $m = 0$. Then, from estimates (iii) and (ii) we have

$$\|x\|_\infty \leq \frac{T}{2}\phi^{-1}\left(2kl + a\frac{T}{2} + \frac{\|e\|_{L^1}}{2}\right) < \frac{T}{2}\phi^{-1}\left(2k\pi + a\frac{T}{2} + \frac{\|e\|_{L^1}}{2}\right) < l,$$

$$\|y\|_\infty \leq kl + a\frac{T}{2} + \frac{\|e\|_{L^1}}{2} < k\pi + a\frac{T}{2} + \frac{\|e\|_{L^1}}{2},$$

and so $(x(t), y(t)) \in \Omega_1$ for all $t \in [0, T]$.

Claim 2. $d_B(f_0, \Omega_1, (0,0)) \neq 0$ *where* $f_0(x, y) = (\phi^{-1}(y - kx), -a\sin(x))$.

Since $0 < l < \pi$ and $a \neq 0$, the only zero of f_0 in $\overline{\Omega_1}$ is $(0,0)$ and then $d_B(f_0, \Omega_1, (0,0))$ is well-defined. Since Ω_1 is symmetric with $(0,0) \in \Omega_1$ and f_0 is continuous in $\overline{\Omega_1}$ and odd, then by Borsuk's theorem it follows that $d_B(f_0, \Omega_1, (0,0))$ is odd. Thus the claim follows.

Finally, the proof of the existence of a T-periodic solution x_2 satisfying (3.23) is analogous by using the open bounded set

$$\Omega_2 = \left\{(x, y) \in \mathbb{R}^2 : |x - \pi| < l,\ |y - k\pi| < k\pi + a\frac{T}{2} + \frac{\|e\|_{L^1}}{2}\right\}.$$

Indeed, the analogous to *Claim 1* follows from the estimates provided in Lemma 3.4 with $m = 1$. On the other hand, for the analogous to *Claim 2*, consider the homeomorphism $T : \Omega_1 \to \Omega_2$ defined by $T(x, y) = (x + \pi, y + k\pi)$. By the Product Formula (see [12, Theorem 5.1]) we have

$$\begin{aligned} d_B(f_0, \Omega_2, (0,0)) &= d_B(f_0 \circ T, \Omega_1, (0,0)) \cdot d_B(T, \Omega_1, z) \\ &= d_B(f_0 \circ T, \Omega_1, (0,0)), \end{aligned}$$

since $d_B(T, \Omega_1, z) = 1$ for any $z \in \Omega_2$. So we can apply again Borsuk's theorem to get $d_B(f_0 \circ T, \Omega_1, (0,0)) \neq 0$ since $(f_0 \circ T)(x, y) = (\phi^{-1}(y - kx), a\sin(x))$ is also a continuous and odd function. □

As a consequence of the previous result, we obtain the following "uniform" estimate.

Corollary 3.1 *If* $cT \leq 2\pi$*, then for any values* $a, k > 0$ *and* $f \in \tilde{C}_T$ *the problem* (3.7) *has at least two geometrically different* T*-periodic solutions* x_1 *and* x_2 *that satisfy* (3.22) *and* (3.23)*, respectively.*

Proof It is enough to realize that

$$\frac{T}{2}\phi^{-1}\left(2k\pi + a\frac{T}{2} + \frac{\|f\|_{L^1}}{2}\right) < \frac{cT}{2},$$

so the assumptions in Theorem 3.5 are satisfied. □

Finally, up to our knowledge, the best quantitative estimate was given by Amster et al. [2]. The idea to obtain their result is to follow the same strategy as in

Theorem 3.4 and take care of the way in which the contradiction is obtained in Step 1 of the proof.

Theorem 3.6 ([2]) *Assume that $cT \leq 2\sqrt{3}\pi$, and there exists $\delta \in \left(0, \frac{\pi}{2}\right)$ such that*

$$\psi(\delta) := 2\delta\cos(\delta) + (cT - 2\delta)\cos\left(\frac{cT}{2\sqrt{3}}\right) > 0.$$

Then, there follow the same conclusions as in Theorem 3.4.

Proof If $cT < \sqrt{3}\pi$, we get the result directly from Theorem 3.4 so we can assume without loss of generality that

$$\sqrt{3}\pi \leq cT \leq 2\sqrt{3}\pi.$$

We proceed as in the proof of Theorem 3.4 by defining

$$\Omega_1 := \left\{x \in C_T^1 : |\overline{x}| < \frac{\pi}{2}, \quad \|x'\|_\infty < c, \quad \|x - \overline{x}\|_\infty < \frac{cT}{2\sqrt{3}}\right\},$$

and considering the homotopic family $\mathcal{M}_\lambda : \overline{\Omega_1} \to C_T^1$ given by (3.16). Now, the only different part with respect to the proof of Theorem 3.4 is in Step 1.

Claim. There is no solution of the equation $x = \mathcal{M}_\lambda x$ for $\lambda \in [0, 1]$ and $x \in \partial\Omega_1$.

As before, if there exist $x \in \partial\Omega_1$ and $\lambda \in [0, 1]$ such that $x = \mathcal{M}_\lambda x$, then should be $|\overline{x}| = \dfrac{\pi}{2}$ and

$$\int_0^T \sin(x(s))ds = 0.$$

Suppose $\overline{x} = \dfrac{\pi}{2}$ (the case $\overline{x} = -\dfrac{\pi}{2}$ is similar), then

$$\left\|x - \frac{\pi}{2}\right\|_\infty < \frac{cT}{2\sqrt{3}} \leq \pi,$$

and hence

$$-\frac{\pi}{2} < x(s) < \frac{3\pi}{2} \quad \text{for all } s \in [0, T].$$

Consequently, defining $A := \dfrac{cT}{2\sqrt{3}} - \dfrac{\pi}{2} > 0$, we have

$$\sin(x(s)) \geq -\sin(A) \quad \text{for all } s \in [0, T],$$

and therefore

$$\begin{aligned}\int_0^T \sin(x(s))ds &= \int_0^T (\sin(x(s)) + \sin(A))ds - \int_0^T \sin(A)ds \\ &\geq \int_{\{s:\, |x(s)-\frac{\pi}{2}|\leq\delta\}} (\sin(x(s)) + \sin(A))\, ds - T\,\sin(A) \\ &\geq \left(\sin\left(\frac{\pi}{2}-\delta\right) + \sin(A)\right)\mu\left(\left\{s:\, |x(s)-\frac{\pi}{2}|\leq\delta\right\}\right) - T\,\sin(A) \\ &\geq \left(\sin\left(\frac{\pi}{2}-\delta\right) + \sin(A)\right)\frac{2\delta}{c} - T\,\sin(A) \\ &= \left(\cos(\delta) - \cos\left(\frac{cT}{2\sqrt{3}}\right)\right)\frac{2\delta}{c} + T\cos\left(\frac{cT}{2\sqrt{3}}\right) = c\psi(\delta) > 0,\end{aligned}$$

attaining in this way a contradiction. Notice that the Lebesgue measure

$$\mu\left(\left\{s:\, |x(s)-\frac{\pi}{2}|\leq\delta\right\}\right) \geq \frac{2\delta}{c},$$

because of $\|x'\|_\infty < c$ and the periodicity of x.

The rest of the proof runs in the same way as that of Theorem 3.4. □

Corollary 3.2 *Assume that* $cT \leq 6.318$. *Then, there follow the same conclusions as in Theorem 3.4.*

Proof As in the previous theorem, it is enough to show the result for

$$\sqrt{3}\pi \leq cT \leq 6.318.$$

Now, a numerical computation (see Fig. 3.1) shows that for $cT = 6.318$ holds that

$$\psi(0.98) > 0.$$

So, the result follows from Theorem 3.6 by taking into account that

$$\psi(\delta) = 2\delta\cos(\delta) + (cT - 2\delta)\cos\left(\frac{cT}{2\sqrt{3}}\right)$$

is decreasing with respect to cT on the interval $[\sqrt{3}\pi, 2\pi]$ for any $\delta \in \left(0, \frac{\pi}{2}\right)$. □

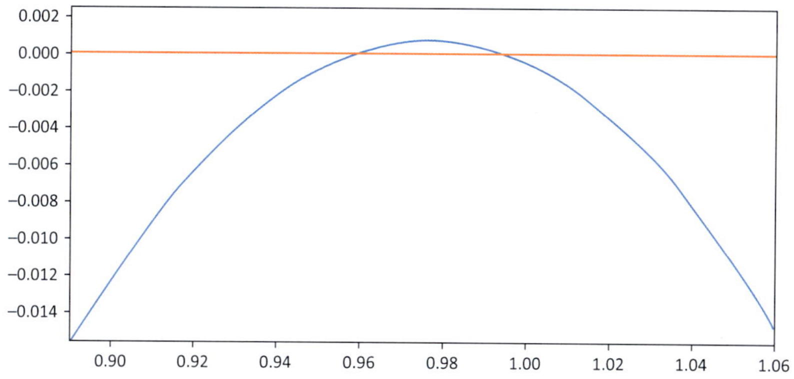

Fig. 3.1 Graphic of $\psi(\delta)$ with $cT = 6.318$

References

1. J.M. Alonso, Nonexistence of periodic solutions for a damped pendulum equation. Differ. Integr. Equ. **10**(6), 1141–1148 (1997)
2. P. Amster, M.P. Kuna, D.P. Santos, On the solvability of the periodically forced relativistic pendulum equation on time scales. Electron. J. Qual. Theory Differ. Equ. **2020**(62), 1–11 (2020)
3. C. Bereanu, P. Jebelean, J. Mawhin, Periodic solutions of pendulum-like perturbations of singular and bounded Φ-Laplacians. J. Dynam. Differ. Equ. **22**, 463–471 (2010)
4. C. Bereanu, J. Mawhin, Existence and multiplicity results for some nonlinear problems with singular ϕ-Laplacian. J. Differ. Equ. **243**(2), 536–557 (2007)
5. C. Bereanu, P.J. Torres, Existence of at least two periodic solutions of the forced relativistic pendulum. Proc. Am. Math. Soc. **140**(8), 2713–2719 (2012)
6. H. Brezis, J. Mawhin, Periodic solutions of the forced relativistic pendulum. Differ. Integr. Equ. **23**, 801–810 (2010)
7. A. Capietto, J. Mawhin, F. Zanolin, Continuation theorems for periodic perturbations of autonomous systems. Trans. Am. Math. Soc. **329**(1), 41–72 (1992)
8. A. Castro, Periodic solutions of the forced pendulum equation, in *Diff. Equations*, ed. by Ahmad, Keener, Lazer (Academic Press, New York, 1980), pp. 149–160
9. J.A. Cid, On the existence of periodic oscillations for pendulum–type equations. Adv. Nonlinear Anal. **10**(1), 121–130 (2021)
10. J.A. Cid, P.J. Torres, On the existence and stability of periodic solutions for pendulum-like equations with friction and ϕ-Laplacian. Discrete Contin. Dyn. Syst. **33**, 141–152 (2013)
11. N. Dancer, On the use of asymptotics in nonlinear boundary value problems. Ann. Mat. Pura Appl. **(4) 131**, 167–185 (1982)
12. K. Deimling, *Nonlinear Functional Analysis* (Springer, 1985)
13. G. Hamel, Ueber erzwungene Schingungen bei endlischen Amplituden. Math. Ann., **86**, 1–13 (1922)
14. J. Mawhin, The forced pendulum equation: a paradigm for nonlinear analysis and dynamical systems. Expo. Math. **6**, 271–287 (1988)
15. J. Mawhin, Seventy-five years of global analysis around the forced pendulum equation. Proc. EQUADIFF **9**, 115–145 (Brno, 1997)
16. J. Mawhin, Global results for the forced pendulum equation, in *Handbook of Differential Equations*, vol. 1 (Elsevier, 2004), pp. 533–589

17. J. Mawhin, Periodic solutions of the forced pendulum: classical vs relativistic. Le Matematiche **LXV**, 97–107 (2010)
18. J. Mawhin, M. Willem, Multiple solutions of the periodic boundary value problem for some forced pendulum-type equations. J. Differ. Equ. **52**, 264–287 (1984)
19. R. Ortega, A counterexample for the damped pendulum equation. Bulletin de la Classe de Sciences, Académie Royale de Belgique **LXXIII**, 405–409 (1987)
20. R. Ortega, E. Serra, M. Tarallo, Non-continuation of the periodic oscillations of a forced pendulum in the presence of friction. Proc. Am. Math. Soc. **128**(9), 2659–2665 (2000)
21. P.J. Torres, Periodic oscillations of the relativistic pendulum with friction. Phys. Lett. A **372**, 6386-6387 (2008)
22. P.J. Torres, Nondegeneracy of the periodically forced Liénard differential equation with ϕ-Laplacian. Commun. Contemp. Math. **13**(2), 283–292 (2011)
23. M. Willem, Oscillations forcées de l'équation du pendule. Publ. IRMA Lille **3**, V-1-V-3 (1981)

Chapter 4
Recent Advances on Periodic Motions in Parallel-Plate Electrostatic Actuators

Andrés Rivera and John A. Arredondo

4.1 Introduction

Microelectromechanical systems, known short as MEMS, refer to every technological device fabricated using a combination of semiconductor materials and mechanical components, usually arranged on a silicon substrate with sizes in microns scale. Its history dates back to 1967, when Harvey C. Nathanson, a physicist and electrical engineer who worked at the RCA laboratories in Princeton, introduced in a research paper *the resonant gate transistor*, the first mathematical model of a MEMS [1]. Nowadays, this model is known as the Nathanson oscillator, and we can consider it, as the first answer to R. Feynman, who asked for the possibility of direct manipulation of individual atoms as a robust form of synthetic chemistry, in a seminal lecture given at the annual American Physical Society meeting of 1959, titled *There's Plenty of Room at the Bottom: An Invitation to Enter a New Field of Physics* [2].

The main purpose of MEMS devices is to control and act on a micro-scale in many technological applications, the most common, sensor-manufacturing, the reason for which investigation in MEMS technology, in the field of physics, engineering, and mathematics, has been increasing significantly during the last decades: to reduce the design of MEMS in time and cost, to develop devices

A. Rivera (✉)
Departamento de Ciencias Naturales y Matemáticas, Pontificia Universidad Javeriana Cali, Facultad de Ingeniería y Ciencias, Cali, Colombia
e-mail: amrivera@javerianacali.edu.co

J. A. Arredondo
Departamento de Matemáticas, Fundación Universitaria Konrad Lorenz, Facultad de Matemáticas e Ingenierías, Bogotá, Colombia
e-mail: alexander.arredondo@konradlorenz.edu.co

P. Amster, P. Benevieri (eds.), *Topological Methods for Delay and Ordinary Differential Equations*, Advances in Mechanics and Mathematics 51,
https://doi.org/10.1007/978-3-031-61337-1_4

of higher quality, to improve their features, optimize the performance of existing devices and design novel devices, in general, to advance technology at the next level. So it is not surprising that since the incursion of the Nathanson model, the number of models that have been developed for MEMS in a variety of scenarios has only increased. But even in very simplified and idealized physical situations, the resulting equations are extremely difficult in any mathematical context. So not only are the applications of true interest but so is the mathematical research of the equations which involves various tools from many different fields of mathematics like nonlinear analysis, qualitative theory of ordinary differential equations, and numerical analysis. The mathematical model originally proposed by Nathanson, for an actuator composed of two parallel plates (electrodes) one fixed and the other moveable, is summarized in the second-order differential equation

$$my''(\tau) + \gamma y'(\tau) + ky(\tau) = f_{\text{ext}}(\tau, y(\tau)) + f_{\text{int}}(\tau, y(\tau)),$$

where $m \in \mathbb{R}^+$ is the mass of a plate actuator, $\gamma, k \in \mathbb{R}^+$ represents the damping coefficient and the spring constant respectively, $y = y(\tau)$ is the displacement at time τ of the moveable plate from its equilibrium position, $f_{\text{ext}}(\tau, y(\tau))$ represents the external force acting on the moveable plate, and finally, $f_{\text{int}}(\tau, y(\tau))$ are the internal forces due to electrostatic and/or piezoelectric effects (more details about this model will be given in Sect. 4.2). Since the appearance of this model, many variations and generalizations of this equation have been carried out, on which from the analytical point of view, the usual techniques of dynamical systems have been applied for the analysis of equilibrium solutions and their stability, see [3, 4] for the details of these facts, and from the point of view of numerical analysis, there have been applied several other techniques for these and similar purposes. For more details, see Chapter 3 in [5] and specifically Section 4 in [6] and the references cited therein.

In the specific setting of periodic oscillations, this document is fully devoted to summarizing some of the most recent results above existence and stability criteria for periodic solutions of equations modeling MEMS, specifically, we use the Nathanson model to illustrate techniques and results used in this frame. The document is organized as follows: In Sect. 4.2 we present the basic ingredients to establish a model for MEMS device, giving the corresponding mathematical description for each of them, until obtaining the general differential equation that governs the Nathanson model and some of its variations. In Sect. 4.3 we show the recent results about the existence, multiplicity, and stability of periodic solutions (with defined sign) for the Nathanson model which corresponds to the study of a Liénard periodic boundary value problem, presenting such an analysis for a graphene-based model, a system under squeeze damping force and a system under delayed controller effects. Two results on the existence and stability of periodic oscillations for the Nathanson model and another parallel-plate actuator known as the Comb-drive finger actuator are presented in Sect. 4.4. Next, in Sect. 4.5 we perform the basic details of the different tools that support the analysis of the models developed in Sect. 4.4. Finally, Sect. 4.6 provides some open mathematical problems in the study of the dynamics of MEMS modeled by Liénard type equations.

4.2 Deduction of a Mathematical Model for MEMS

Following Chapter 3 in [4] as a guideline, below we briefly describe some of the main elements that appear in any mathematical model for MEMS, which together provide the differential equation that governs the motion of an idealized electrostatically device known as the *electrostatic parallel plate actuators*.

Let us start by mentioning that MEMS could have several configurations, one of the most popular designs is those where the device has two parallel capacitor plates (electrodes), one fixed and the other able to move toward or away from the fixed capacitor. Further, the plates could have any form, but for simplicity, it is usual to consider a rectangular shape (see Fig. 4.1). In this case, the attractive forces between the two parallel plates (with opposites charge) are given by Coulomb's law, which depicts the reciprocal force between two charges separated by a certain distance. Additionally, the motion of the movable plate is restricted by a spring system. Under this scheme, we have the following characteristics:

†) The whole system remains at rest when a voltage difference between the electrodes is not applied, then, an electrostatic force is zero, the position called the rest gap distance.

†) As the voltage increased, the capacitor is charged according to $Q = C(\eta)\hat{V}$, where Q is the electrical charge of the capacitor, C is its capacitance, and η is a variable representing the degree of freedom of the movable electrode of the capacitor. With this, the potential energy stored in the capacitor is given by

$$E_c = \frac{1}{2}C(\eta)\hat{V}^2.$$

†) As soon the top electrode moves a small distance toward the bottom electrode while the voltage across the electrodes remains constant, the actuator will gain energy and the power supply will give up internal energy. Therefore the energy generated from the battery must be taken into account. Then, if the initial energy of the battery that is hooked-up to the capacitor is assumed to be E_0. When connected to the capacitor, the battery provides the electrical charge Q and an electrical potential $-Q\hat{V}$, which is negative since it is lost from the battery.

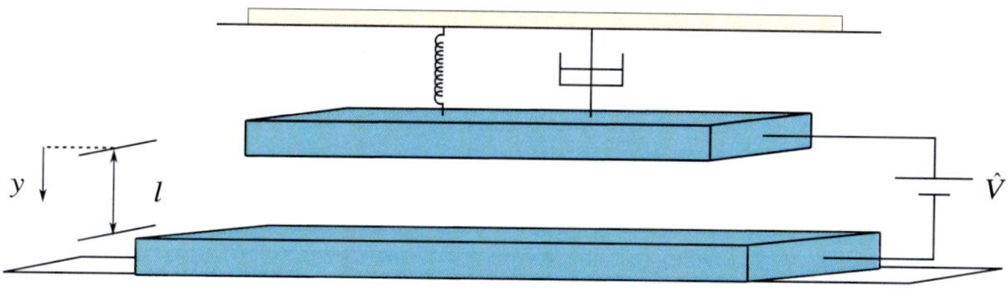

Fig. 4.1 Schematic drawing of a two-parallel plate capacitor. The plates are initially separated at a distance l. The lower plate is fixed, while the upper plate moves a distance y under the effects of a spring force, a damping force, and an electrostatic force

Hence, the new potential energy of the battery denoted E_b, is reduced to

$$E_b = E_0 - C(\eta)\hat{V}^2.$$

†) The force between the two electrodes of the capacitor is given by

$$f_c = \frac{-\partial E(\eta)}{\partial \eta} = \frac{1}{2}\frac{\partial C(\eta)}{\partial \eta}\hat{V}^2.$$

For electrodes of simple geometries analytical expression for $C(\eta)$ can be found, but in more general cases numerical approximation techniques must be used to characterize the dependence of $C(\eta)$ on the displacement η. The most simple situation could be a capacitor formed by two parallel rectangular plates, one of them is kept stationary while the other is allowed to move toward the stationary electrode only, as is shown in Fig. 4.1. In this case, $\eta = y$, where y is the distance between the two electrodes. The physical dimensions of the electrodes, i.e., the width b and length L are the same for both for manufacturing reasons. Solving the Laplace partial equation that governs the electrostatic interaction between two rectangular plates, the capacitance of the capacitor can be expressed as

$$C(y) = \varepsilon_r \varepsilon_o \frac{Lb}{y},$$

where Lb is the overlap area between the electrodes, ε_r is the relative permittivity of the gap space medium concerning the free space and ε_o is the dielectric constant in free space. Under this setting, the force between two rectangular electrodes is given by

$$f_c = -\frac{\varepsilon_r \varepsilon_o A}{2y^2}\hat{V}^2,$$

where the minus sign indicates that the force on the movable electrode is of an opposite sign to the positive y-direction.

Remark 4.1 In MEMS design the movable electrode is a flexible structure, hence, when deflected a distance y due to the electrostatic force, the natural resorting force of the structure opposes the electrostatic force. For a low electrostatic force, the equilibrium between the restoring and electrostatic force is achieved. Nevertheless, when the voltage increased too much, electrostatic force overcomes the resorting force of the structure until it collides with the stationary electrode. This phenomenon is known as *pull-in instability*.

Remark 4.2 Another prevalent capacitor form consists of a comb-drive structure (simply a brush system of capacitors) consisting of multiple interdigitated or non-interdigitated fingers (see Fig. 4.2). This is a system of two comb structures (or more), one stationary and the other one movable connected to a system of springs or

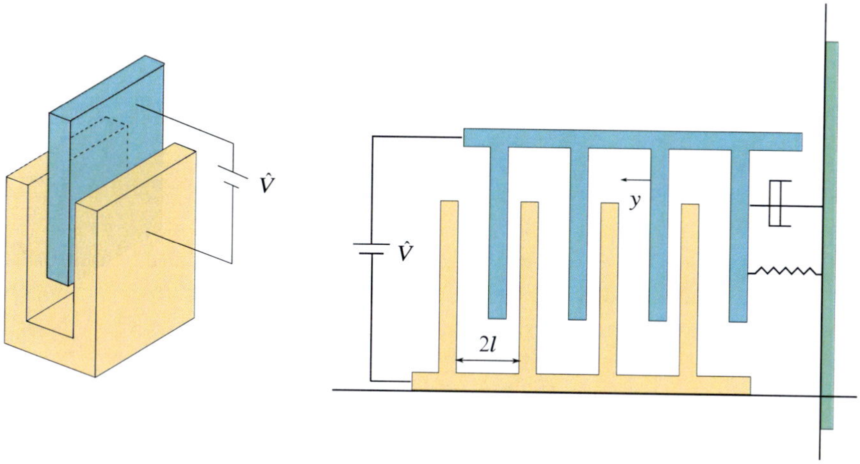

Fig. 4.2 Schematic graphic of a Comb-drive parallel plate actuator, a brush system of capacitors, consisting of multiple interdigitated fingers

similar flexible structures. Typically the teeth are arranged so that they can slide past one another until each tooth occupies the slot in the opposite comb. The stationary and movable fingers form parallel plate capacitors among each other. If the capacitors are biased by a voltage $\hat{V}$, then two electrostatic forces will be generated on the movable finger. One acts on its left side $f_{c_{\text{left}}}$ in the positive y-direction and an opposite one acts on its right side $f_{c_{\text{right}}}$.

$$f_{c_{\text{left}}} = \frac{\varepsilon(lt)\hat{V}^2}{2(l-y)^2}, \quad \text{and} \quad f_{c_{\text{right}}} = \frac{\varepsilon(lt)\hat{V}^2}{2(l+y)^2}.$$

If $y = 0$, the electrostatic forces acting on the movable finger cancel each other. If $y \geq 0$, a resultant force $f_{c_{\text{right}}} - f_{c_{\text{left}}}$ will act on the movable finger. There is a limit on the maximum stable displacement of the comb before the electrostatic force overcomes the elastic resorting force from the tethers before it snaps, *pull-in*, to the stationary combs.

Damping Mechanism Perhaps, one of the most important features related to the fabrication, design, and performance of electrostatic MEMS is energy dissipation. This aspect can be divided into external dissipation and internal dissipation. The first one is related to the environment of operability of the device, and the second one considers the material properties of the device itself. For some MEMS devices, the external dissipation mechanism can be modeled by a damping force $f_D(y, y')$, that is proportional to the speed $y' = y'(\tau)$ of the movable part. Several types of damping mechanisms can be used in MEMS, depending on the specific application and requirements, for the two best known, viscous damping and squeeze-film damping, the corresponding damping coefficient is given respectively by

$$a) \quad \tilde{c}_D(y) = c, \quad \text{and} \quad b) \quad \tilde{c}_D(y) = \frac{\gamma}{\vartheta(y)^3}, \quad c, \gamma \in \mathbb{R}^+.$$

- Case a) refers to a linear damping force, where the parameter c is the viscosity coefficient of the fluid in which the device is placed. In fact, this constant can be expressed as $c = \dfrac{6\pi\mu\tilde{r}^2}{\sigma l}$ where μ is the dynamic viscosity of the fluid, $\tilde{r}$ is the radius of the gap, σ is the density of the fluid, and l is the length of the gap.
- Case b) refers to a nonlinear damping force called squeeze-film damping that appears on devices with a big surface that encloses a fluid in a small area. The parameter γ depends on the surface area and the viscosity coefficient of the fluid, and $\vartheta(y)$ is the variable distance between the close surfaces.

Other types of damping that are used and have been studied in MEMS literature are:

- Solid-state damping: This type of damping is due to the internal friction or damping of the material itself. It can be achieved by using materials with high damping coefficients, such as polymers, ceramics, or composites, or by introducing additional damping layers or coatings.
- Electromechanical damping: This type of damping is achieved by applying an electric field to the device, which induces a force that opposes the motion of the mechanical structure. Electromechanical damping is usually used in devices with piezoelectric or electrostatic actuators.
- Active feedback damping: This type of damping is achieved by applying a feedback control signal to the device, which adjusts the driving force or actuation voltage in real time to maintain the desired amplitude or frequency of oscillation. Active feedback damping can be more effective than passive damping mechanisms in devices with nonlinear or time-varying dynamics.

Delayed Feedback Control In electrostatic actuators, time-delay phenomena are inherent to the device or generated by design. To correct this, Delayed Feedback Control (DFC) is a type of control method commonly used in MEMS devices to stabilize oscillations and improve performance. DFC works by introducing a delay between the measurement of a signal and the application of a control action, which can help to suppress any unwanted oscillations or noise in the system. To apply DFC to a MEMS device, a portion of the oscillation signal is fed back into the control system after a certain time delay (Fig. 4.3). This delayed feedback signal is then used to adjust the control signal applied to the device, effectively stabilizing the oscillations and reducing any unwanted noise or interference.

One of the first scientists that applied time-delay feedback controllers was K. Pyragas [7], where the delay was used to stabilize the unstable periodic orbits of a nonlinear chaotic dynamical system by applying a small time-delayed feedback signal. The output signal of this controller is a delayed value of the system output subtracted from it the current output of the system. The delayed time in the Pyragas method has been investigated in several cases, for example, in the works of M.I.

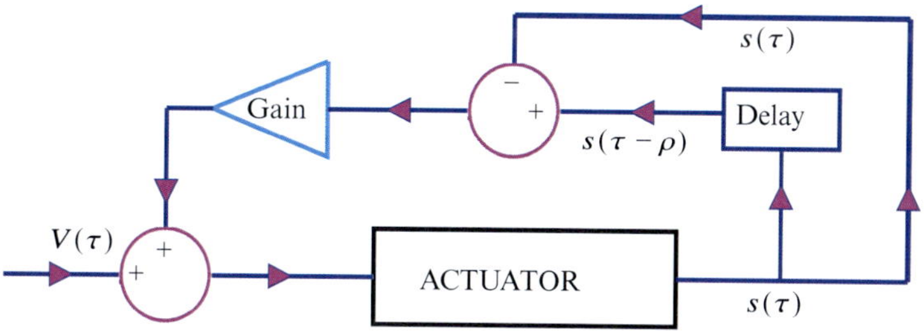

Fig. 4.3 Schematic representation of the delayed feedback controller introduced by K. Pyragas [7]. The control acts over the output signal $s(\tau)$, which could be the system deflection $s(\tau) = y(\tau)$, or the system velocity $s(\tau) = y'(\tau)$. The objective is to measure $\Gamma(s(\tau - \rho) - s(\tau))$ where Γ is the gain controller

Younis (with Nayfeh, Alsaleem, Shao, and Masri) [8–10] describe the dynamics of a resonant microbeam excited electrically, the dynamics of delayed feedback MEMS resonators and the solution for a single degree-freedom resonator model actuated by a DC-voltage source and an AC-voltage source with a delay feedback controller. For this purpose, perturbation, multiple scales method combined with shooting technique and basin of attraction analysis are employed.

In all the referred works, some analytical support is presented after verifying the results experimentally. In feedback controllers with time delay, the output signal of this type of controller is a delayed value of the system output from which the current output of the system is subtracted (see Fig. 4.3). The delay effect is added to the system by adjusting the voltage signal. Position feedback $y_\rho(\tau)$ (see [9]) or velocity feedback $y'_\rho(\tau)$ (see [10]) given by

$$y_\rho(\tau) = y(\tau - \rho) \qquad \text{and} \qquad y'_\rho(\tau) = \frac{d}{d\tau} y(\tau - \rho),$$

respectively, can be used with the delayed time $\rho > 0$. If we combine those situations, the voltage load $\hat{\mathcal{V}}$ can be expressed by

$$\hat{\mathcal{V}}(\tau, \sigma_1(y, y_\rho, \Gamma_1), \sigma_2(y', y'_\rho, \Gamma_2)) = \hat{V}(\tau) + \sigma_1(y, y_\rho, \Gamma_1) + \sigma_2(y', y'_\rho, \Gamma_2),$$

with

$$\sigma_1(y(\tau), y_\rho(\tau), \Gamma_1) = \Gamma_1(y_\rho(\tau) - y(\tau)),$$

and

$$\sigma_2(y'(\tau), y'_\rho(\tau), \Gamma_2) = \Gamma_2(y'_\rho(\tau) - y'(\tau)),$$

where $\Gamma_i \in \mathbb{R}$, $i = 1, 2$, are constants corresponding to the *gains* of the controller with respect to the delayed position and delayed velocity. The gain in a controller for MEMS refers to the amplification factor applied to the error signal or feedback signal to generate the control signal. In other words, the gain determines how much the controller responds to changes in the feedback signal. In addition, the gain can be adjusted to tune the performance of the controller, such as the settling time or steady-state error of the system. In the context of a delayed feedback controller, the gain determines the strength of the feedback signal that is fed back to the system after a certain time delay. A higher gain results in a stronger feedback signal, which can help stabilize the system or suppress oscillations. Finally, a gain constant could be positive or negative, a value that affects the dynamics with which the delay influences the device.

The Differential Equations Let us consider the Nathanson model and the Comb-drive acting without the presence of any controller (Figs. 4.1 and 4.2). Then from Newton's law, the physical forces that intervene in MEMS devices that we have previously described, provides a general equation of a parallel plate actuator is given by the following nonlinear differential equation

$$my''(\tau) = -f_D(y(\tau), y'(\tau)) - f_R(y(\tau)) + f_C(\tau, y(\tau)). \qquad (*)$$

For the Nathanson model, the electrostatic force is given by

$$f_{C,nat}(\tau, y) = \frac{\epsilon A\, \hat{V}^2(\tau)}{2(l-y)^2},$$

where l is the fixed distance between the movable electrode and the fixed one. Meanwhile, for the Comb-drive model, the electrostatic force is

$$f_{C,comb}(\tau, y) = \frac{\epsilon A \hat{V}^2(\tau)}{2(l - y(\tau))^2} - \frac{\epsilon A \hat{V}^2(\tau)}{2(l + y(\tau))^2}.$$

Let us assume that the moveable electrode is attached to a linear (or nonlinear) spring in both models. Therefore, the restoring force can be expressed by $f_R(y) = ky + r(y)y$ with $r(y)$ an even function, and $k > 0$ is the linear stiffness coefficient of the spring. Then, under the change of variables

$$x = \frac{y}{l}, \quad \text{and} \quad t = \omega_0 \tau, \quad \omega_0 = \sqrt{k/m},$$

the equation $(*)$ for the Nathanson model becomes

$$\ddot{x}(t) + c(x(t))\dot{x}(t) + q(x(t)) = \frac{eV^2(t)}{(1 - x(t))^2}, \qquad (4.1)$$

where

$$c(x) = \frac{\tilde{c}_D(lx)}{\sqrt{mk}}, \quad q(x) = x + \frac{r(lx)x}{k}, \quad V(t) = \hat{V}(t/\omega_0), \quad \text{and} \quad e = \frac{\epsilon A}{2kl^3}.$$

Analogously, for the Comb-drive finger model, the equation of motion is given by

$$\ddot{x}(t) + c(x(t))\dot{x}(t) + q(x(t)) = \frac{4ex(t)V^2(t)}{(1 - x^2(t))^2}. \tag{4.2}$$

Now, if we consider that both actuators act under the presence of a time-delayed controller over the voltage load, then in the new variables the voltage source satisfies

$$V(t) \rightarrow V(t) + \sigma(s(t), s_d(t), g), \quad \text{with} \quad \sigma(s(t), s_d(t), g) = g(s_d(t) - s(t)),$$

with

$$d = \omega_0/\rho, \quad \text{and} \quad g = \begin{cases} l\Gamma, & \text{if} \quad s(\tau) = y(\tau), \\ l\omega_0\Gamma, & \text{if} \quad s(\tau) = y'(\tau). \end{cases}$$

Therefore, the time-delayed differential equation for the Nathanson model under a feedback controller is given by

$$\ddot{x}(t) + c(x(t))\dot{x}(t) + q(x(t)) = \frac{e[V(t) + \sigma(s(t), s_d(t), g)]^2}{(1 - x(t))^2}, \tag{4.3}$$

and for the Comb-drive actuator is

$$\ddot{x}(t) + c(x(t))\dot{x}(t) + q(x(t)) = \frac{4ex(t)[V(t) + \sigma(s(t), s_d(t), g)]^2}{(1 - x^2(t))^2}. \tag{4.4}$$

In the following sections, we show some recent results on periodic solutions of the Liénard type Eqs. (4.1) and (4.2) and for the time-delayed Liénard equation (4.3).

4.3 Existence and Stability of Periodic Motions with a Defined Sign

The theoretical framework where we can place almost all the recent papers [11–17] devoted to the study of periodic motions for the parallel actuators (4.1) and (4.2) corresponds to the study of the periodic boundary value problem

$$\begin{aligned} &\ddot{x}(t) + F(t, x(t), \dot{x}(t)) = 0, \\ &x(0) = x(T), \quad \dot{x}(0) = \dot{x}(T), \end{aligned} \tag{4.5}$$

with $F : \Omega \to \mathbb{R}$ a continuous function, periodic in time with period T, i.e.,

$$F(t+T, x, \dot{x}) = F(t, x, \dot{x}), \quad \forall (t, x, \dot{x}) \in \Omega,$$

where $\Omega \subseteq \mathbb{R}\times]x_1, x_2[\times\mathbb{R}$ is an open connected set with $-\infty \leq x_1 < x_2 \leq \infty$.

The Liénard periodic boundary value problem given by

$$\begin{aligned} &\ddot{x}(t) + c(x(t))\dot{x}(t) + Q(t, x(t)) = 0, \\ &x(0) = x(T), \quad \dot{x}(0) = \dot{x}(T), \end{aligned} \tag{4.6}$$

is an example of (4.5) which is commonly used to describe the dynamics of several mechanical oscillators, and in particular, the motions of MEMS with one degree of freedom [4, 18]. Therefore, the search conditions for the existence of solutions of (4.5) have always been an open and interesting matter.

Remark 4.3 Direct computations prove that any solution $x = x(t)$ of (4.5) is a T-periodic function, i.e., $x(t+T) = x(t)$ for all $t \in \mathbb{R}$.

One of the most well-known methods that give sufficient conditions for the existence of solutions of (4.5) is the *method of upper and lower solutions*, introduced by E. Picard in 1893 [19] and developed by G. Scorza Dragoni in 1931 [20]. In general terms, the idea behind the method is two find a pair of functions $\xi(t)$ and $\eta(t)$ (*called upper and lower solutions*) satisfying the differential inequalities

$$\ddot{\xi}(t) + F(t, \xi(t), \dot{\xi}(t)) \leq 0, \quad \text{and} \quad \ddot{\eta}(t) + F(t, \eta(t), \dot{\eta}(t)) \geq 0, \quad \forall t \in [0, T]$$

an appropriate boundary conditions (see Definition 4.1 in Sect. 4.5) respectively, such that $x_1 \leq \eta(t) \leq \xi(t) \leq x_2$ for all $t \in [0, T]$. We may think of these functions as two "fences." Then under appropriate conditions, the method of upper and lower solutions provides sufficient conditions for the existence of at least one solution of (4.5) in the "funnel"

$$E_{\eta,\xi} = \{(t, x, \dot{x}) \in [0, T]\times]x_1, x_2[\times \mathbb{R} : \eta(t) \leq x \leq \xi(t)\}.$$

If the fences satisfy $\xi(t) \leq \eta(t)$ for all $t \in [0, T]$ and some additional conditions, the same conclusion holds now on the funnel $E_{\xi,\eta}$. In Sect. 4.5, we present a brief introduction to this method, adapted to the boundary value problem (4.5).

Once the existence of an isolated periodic solution $\varphi = \varphi(t)$ of the Liénard periodic problem (4.6) when $c(x) = c$, $c \in \mathbb{R}^+$, is guaranteed, questions concerning the stability properties of φ could be solved using the topological invariant $\mathrm{ind}(\varphi)$ known as *index of the periodic solution* φ. These beautiful connections between the index of φ and their stability properties can be found in [21–24].

From now on we assume the following conditions over the function $V(t)$:

C1. $V(t)$ is positive for all $t \in \mathbb{R}$,

C2. $V(t) = V_0 + \delta v(t)$, where $V_0, \delta \in \mathbb{R}^+$, $v \in C(\mathbb{R}/T\mathbb{Z})$ for some $T > 0$ and with zero average, i.e., $\int_0^T v(t)\,dt = 0$,

and define the quantities

$$V_m := \min_{t\in\mathbb{R}} V(t), \quad \text{and} \quad V_M := \max_{t\in\mathbb{R}} V(t).$$

Let us start by considering the nonlinear oscillator (4.1) under constant damping coefficient effects and linear restoring force, i.e.,

$$\ddot{x}(t) + \varrho \dot{x}(t) + x(t) = \frac{eV^2(t)}{(1 - x(t))^2}, \tag{4.7}$$

where $\varrho = c/\sqrt{mk}$. Best of our knowledge, in 2013 A. Gutierrez and P. Torres in [14] presented for the first time a rigorous mathematical study of positive periodic motions of (4.7) using the method of upper and lower solutions. Moreover, in addition to Topological Degree Theory, the connection between the stability and index of a periodic solution the authors prove a saddle-node bifurcation of positive periodic solutions.

The existence result dictates the following:

Theorem 4.1 (Theorem 15 [14]) *There exists $e_0 > 0$ such that:*

1. *If $e > e_0$, then* (4.7) *has no positive T-periodic solutions.*
2. *If $e = e_0$ then* (4.7) *has at least one positive T-periodic solution.*
3. *If $e < e_0$ then* (4.7) *has at least two positive T-periodic solutions.*

Furthermore, e_0 satisfies

$$\frac{4}{27V_M^2} \le e_0 \le \frac{4}{27V_m^2}. \tag{4.8}$$

Remark 4.4 For a given input voltage source $V(t)$, the conditions (4.8) provide a type of control over the physical dimensions of fabrication of the device in terms of V_m and V_M. Moreover, if $V(t) = V_0$, $\forall t \in \mathbb{R}$, and it is assume that k, A are fixed, then the conditions (4.8) are equivalent to

$$\frac{\epsilon A}{2kl_0^3} = \frac{4}{27V_0^2}, \quad \Leftrightarrow \quad V_0 = \sqrt{\frac{8kl_0^3}{27\epsilon A}} =: V_{pull},$$

leading to the theoretical critical DC-voltage value V_{pull} (known as *static pull-in voltage*) where the pull-in instability phenomena emerge. This phenomenon produces mechanical failures in the device. We refer to [4, 18] and the references therein for a more comprehensive study of the pull-in instability effects in MEMS.

The idea behind the proof of Theorem 4.1 consists in showing the existence of two pair of constants $\eta, \xi \in [0, 1[$, such that the functions $x(t) = \xi$, and $x(t) = \eta$, are constant upper and lower solutions of (4.7) respectively. Then, apply Theorem 4.10 if the upper and lower solutions are well-ordered, i.e., $\eta < \xi$, or apply Theorem 4.11 if they are in reverse order, i.e., $\xi < \eta$ (see Sect. 4.4). This methodology proves the existence of at least one periodic solution of (4.7) lying in $[\eta, \xi]$ or $[\xi, \eta]$ in the corresponding case. Moreover, the constants ξ and η satisfy the equations

$$\phi(x) = eV_m^2, \quad \text{and} \quad \phi(x) = eV_M^2, \quad \text{with} \quad \phi(x) = x(1-x)^2, \tag{4.9}$$

respectively. Therefore, for a given voltage load $V(t)$, the authors in [14] consider two parametric equations

$$\phi_m(x, e) = 0, \quad \text{and} \quad \phi_M(x, e) = 0,$$

with $\phi_m(x, e) = \phi(x) - eV_m^2$ and $\phi_M(x, e) = \phi(x) - eV_M^2$.

This methodology works in the same way if instead of considering e as a control parameter, we think as parameters the values V_m and V_M, i.e., for fixed dimensions of the device, we appropriately control the input voltage load. A schematic representation of this methodology is given in Fig. 4.4 for the case of the Nathanson model based on graphene material. This is precisely the core of the

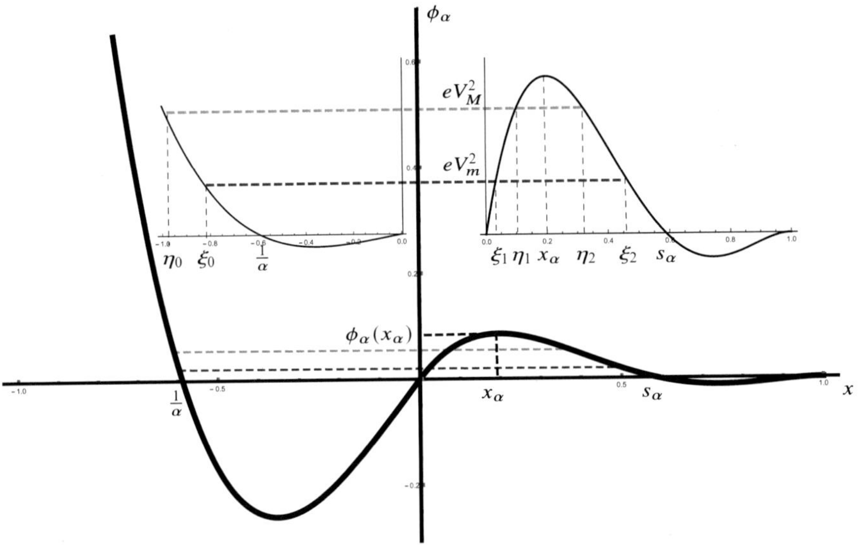

Fig. 4.4 Graph of $\phi_\alpha(x) = x(1-\alpha|x|)(1-x)^2$ for $\alpha > 1$. Since $\phi_\alpha(x) > 0$ for all $x \in]-\infty, 1/\alpha[$ the method of upper and lower solutions provides the existence of at least one negative periodic solution of (4.7). The left (right) inserted graph is a zoom of $\phi_\alpha(x)$ for $x \in [-1, 0]$ ($x \in [0, 1]$) and the constant upper and lower solutions ξ_0, η_0 ($\xi_i, \eta_i, i = 1, 2$)

approach found in [11–13, 15–17] concerning the existence problem of periodic solutions.

4.3.1 Graphene-Based Nathanson Model

With the constant aim of manufacturing and designing devices of higher quality and better performance, the advent of novel materials allows researchers in MEMS to carry out experimental and theoretical models using these new materials. For example, graphene is an appropriate material for designing MEMS due to its high mechanical strength, excellent electrical and thermal conductivity, and high surface area. Graphene is a two-dimensional material consisting of a single layer of carbon atoms arranged in a hexagonal lattice. It is one of the thinnest and strongest materials currently discovered.

The most common theoretical model to test the performance of MEMS-based in graphene is precisely the Nathanson model, for which in [15] the authors provide its first mathematical formulation. In suitable units, the restoring force of the non-dimensional Nathanson graphene-based actuator is given by

$$q_\alpha(x) = x - \alpha|x|x,$$

with α a positive parameter that relates the graphene elastic properties and the device's dimensions. Therefore, the Nathanson model based on graphene materials and an under constant damping coefficient is now given by

$$\ddot{x}(t) + \mu\dot{x}(t) + x - \alpha|x(t)|x(t) = \frac{eV^2(t)}{(1 - x(t))^2}, \tag{4.10}$$

with $\alpha, \mu > 0$. In [25] the authors provide a numerical and analytical study of periodic solutions of (4.10). In particular, considering the function

$$\phi_\alpha(x) = q_\alpha(x)(1 - x)^2.$$

In [15, 16] we found the following properties of $\phi_\alpha(x)$, for each $\alpha > 0$,

1. ϕ_α has roots in $x = 0$, $x = \pm 1/\alpha$ and $x = 1$.
2. Let $s_\alpha = \min\{1, 1/\alpha\}$. Then, $\phi_\alpha \in C^1(\mathbb{R}) \cup C^2(]0, s_\alpha[)$ and $\phi_\alpha > 0$ on $]0, s_\alpha[$.
3. $\phi_\alpha(x)$ has in $]-\infty, 0[$ exactly one negative local minimum and one positive local maximum in $[0, s_\alpha[$. Moreover, if $x_\alpha \in [0, s_\alpha[$ is the unique critical point, it follows that $\phi'_\alpha(x) > 0$, $\forall x \in]0, x_\alpha[$ and $\phi'_\alpha(x) < 0$, $\forall x \in]x_\alpha, s_\alpha[$. Therefore,

$$\phi_\alpha(x_\alpha) = \max_{[0, s_\alpha]} \phi_\alpha(x).$$

As we mentioned, the authors in [16] follow a similar strategy to the one presented in [14] by starting with the search for constant upper and constant lower solutions, as solutions of the equations

$$\phi_\alpha(x) = eV_m^2, \quad \text{and} \quad \phi_\alpha(x) = eV_M^2, \tag{4.11}$$

respectively. Next, we presented an adapted version of Theorem 1 in [16].

Theorem 4.2 (Theorem 1 [16]) *Let be* ξ_1, η_1 *constant upper and lower solutions of* (4.10) *given as solutions of* (4.11) *respectively and such that* $0 < \xi_1 < \eta_1 < x_\alpha$. *Suppose that* $eV_M^2 < \phi_\alpha(x_\alpha)$, $0 < T < \frac{\pi(1-x_\alpha)}{\sqrt{\phi'_\alpha(\xi_1)}}$ *and* $0 < \mu < \mu^*$ *where* μ^* *is given by*

$$\mu^* = \min\left\{2\frac{\sqrt{\phi'_\alpha(\eta_1)}}{(1-\eta_1)}, \mu(\iota^*)\right\}$$

with

$$\mu(\iota^*) := \max_{m\in[\iota^*, \pi^2/T^2]} \frac{m-s}{\sqrt{m}} \cot\left(\sqrt{m}\frac{T}{2}\right), \qquad \iota^* := \frac{\phi'_\alpha(\xi_1)}{(1-\xi_1)}.$$

Then, the Eq. (4.10) *admits at least three* T*-periodic solutions* $\varphi_0(t)$, $\varphi_1(t)$ *and* $\varphi_2(t)$ *such that*

$$\varphi_0(t) < 0 < \varphi_1(t) < \varphi_2(t),$$

moreover,

1. $\varphi_1(t)$ *locally asymptotically stable and* $\varphi_0(t)$ *is unstable.*
2. *Assume that*

$$1 - 2eV_m^2 \leq \left(\frac{\pi}{T}\right)^2 + \frac{\mu^2}{4}.$$

Then, $\varphi_1(t)$ *and* $\varphi_2(t)$ *are the only positive* T*-periodic solutions of* (4.10) *and* $\varphi_2(t)$ *is unstable.*

Finally, if $eV_M^2 > \phi_\alpha(x_\alpha) \geq eV_m^2$ *or* $eV_m^2 > \phi_\alpha(x_\alpha)$. *Then there exists exactly a negative* T*-periodic solution* $\hat{\varphi}_0(t)$ *of* (4.10) *which is unstable.*

It is worth highlighting that if $\alpha > 0$, there is always a pair of constant upper and lower solutions ξ, η respectively, and almost three pairs of constant upper and lower solutions if $eV_M^2 < \phi_\alpha(x_\alpha)$, in contrast with the Eq. (4.10) for which there are almost two pairs of constant upper and lower solutions. Furthermore, if $eV_m^2 > \phi_\alpha(x_\alpha)$ then (4.10) do not admit positive T-periodic solutions. In consequence, the assumption

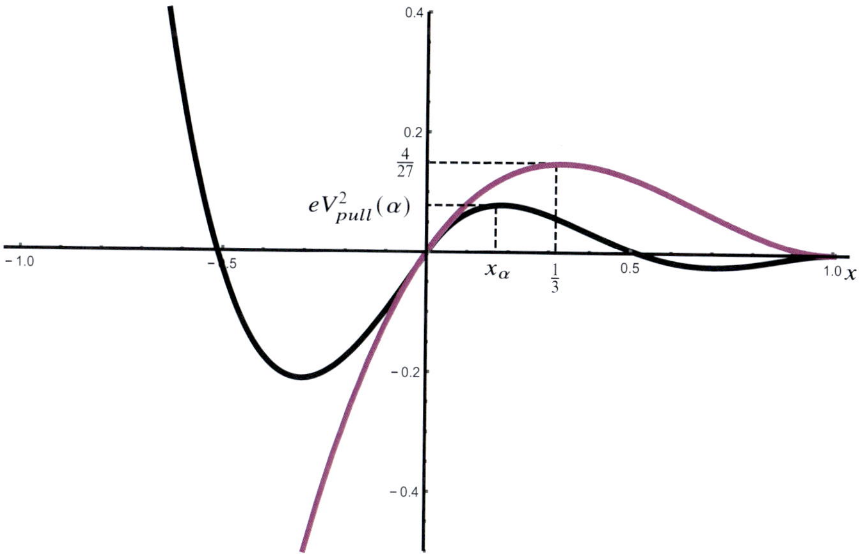

Fig. 4.5 Graph of $\phi_\alpha(x) = x(1-\alpha|x|)(1-x)^2$ for $\alpha > 1$ (Black) and graph of $\phi(x) = x(1-x)^2$ (Magenta). From the graphs, it can be seeing the analytical result $eV^2_{pull}(\alpha) < eV^2_{pull}(0) = 4/27$

$eV_m^2 > \phi_\alpha(x_\alpha)$ and $eV_M^2 < \phi_\alpha(x_\alpha)$ in Theorem 4.2 resembles the assumptions *1.* and *3.* in Theorem 4.1 respectively (Fig. 4.5).

Remark 4.5 In [16, 25] can be found the analytical study of the existence of periodic solutions for a more general Liénard equation given by

$$\ddot{x} + \mu\dot{x} + h(x) = \frac{G(t)}{(1-x)^2},$$

with $\mu > 0$, $h(x)$ in a special class of odd real value functions, and $G(t)$ a positive periodic function. This covers some variations of the mechanical properties of the Nathanson model. For example, linear or nonlinear restoring force, based or not on graphene materials.

In addition, in [25] analyzes the corresponding static pull-in voltage $V_{pull}(\alpha)$ for the graphene-based Nathanson model, which in this case is given by the monotone decreasing function

$$V_{pull}(\alpha) = \sqrt{\frac{(2\alpha + 3 - a_\alpha)(5 - 2\alpha + a_\alpha)(6\alpha - 3 + a_\alpha)^2}{4096\alpha^3 e}},$$

with $a_\alpha = \sqrt{4\alpha^2 - 4\alpha + 9}$. In the limit case $\alpha \to 0^+$ is $V_{pull}(0) = \sqrt{\frac{4}{27e}}$, precisely the static pull-in voltage of the classical Nathanson model (see Remark 4.2). It is no difficult to see that $V_{pull}(\alpha) = \sqrt{\phi_\alpha(x_\alpha)/e}$. Finally, [16] provides numerical computation on stability and basin of attraction associated with the positive T-periodic solutions given in Theorem 4.2 using the stroboscopic Poincaré map, showing that under appropriate control parameters, the region of initial conditions where the actuator operates safety (avoiding possible pull-in stability effects) for the canonical Nathanson model is smaller than for the graphene-based model when both operate in the same mechanical and structural conditions. We may say that this corresponds to the decreasing monotony property of $V_{pull}(\alpha)$.

4.3.2 Periodic Solutions Under Squeeze Damping Force

As we mentioned in the introduction, one of the most common damping forces in parallel plate actuators is the called *squeeze damping force*. In this case, for the Nathanson model, the equation of motion of the moveable plate given by (4.1) can be written as

$$\ddot{x} + \frac{\gamma}{(1 - x(t))^3}\dot{x}(t) + q(x(t)) = \frac{eV^2(t)}{(1 - x(t))^2}. \tag{4.12}$$

Despite the several papers devoted to periodic motions in actuators under a linear damping force, there are few analytical results concerning the existence of periodic solutions of (4.12). To the best of our knowledge [13] is one of the first papers devoted to this problem, where the author's approach is throughout Theorem 4.10 and Theorem 4.11. An adapted version of the main result in [13] with respect to periodic solutions of (4.12) is presented below.

Theorem 4.3 (Theorem 7 [13]) *Let us assume $eV_M^2 \leq 4/27$ and $q(x) = x$. Then there exist constants upper and lower solutions ξ_i, η_i, $i = 1, 2$ of* (4.12) *given as solutions of* (4.9) *respectively and such that*

$$0 < \xi_1 < \eta_1 \leq 1/3, \quad \textit{and} \quad 1/3 \leq \eta_2 < \xi_2 < 1.$$

Furthermore,

1. *Equation* (4.12) *admits a T-periodic solution $\psi_2(t)$ such that*

$$\eta_2 \leq \psi_2(t) \leq \xi_2 \qquad \forall t \in [0, T].$$

2. *Let*

$$N = \frac{\gamma}{(1 - \eta_1)^3}, \quad \hat{a} = \max\left\{N, \eta_1 - \xi_1\right\}, \quad M = \frac{3\gamma R_*}{(1 - \eta_1)^4} + \frac{1 - 3\xi_1}{1 - \xi_1},$$

with $R_* > 0$ the unique solution of

$$R - \ln(R+1) = \hat{a}(\eta_1 - \xi_1).$$

If the following conditions hold

$$M \leq (\pi/T)^2 \quad \text{and} \quad N \leq H(L_*) = \frac{L_* - M}{\sqrt{L_*}} \cot\left(\frac{T\sqrt{L_*}}{2}\right),$$

with $L_* \in [M, (\pi/T)^2]$ *the unique value that satisfies*

$$\sin(T\sqrt{L_*}) = T\sqrt{L_*}\left(\frac{L_* - M}{L_* + M}\right).$$

Then, there exists a periodic solution $\psi_1(t)$ *of* (4.12) *such that*

$$\xi_1 \leq \psi_1(t) \leq \eta_1, \qquad \forall t \in [0, T].$$

Concerning the stability of the periodic solutions $\psi_1(t)$ and $\psi_2(t)$, of (4.12) given by Theorem 4.12, in [13] we found the following result.

Proposition 4.1 *Under the assumption of Theorem 4.3, any possible T-periodic solution of* (4.12) *located* $[\eta_1, \zeta_1]$ *is locally asymptotically stable if*

$$\frac{1 - 3\eta_1}{1 - \eta_1} > \frac{\gamma}{2(1-\eta_1)^4}\left(3R_* + \frac{\gamma}{2(1-\eta_1)^2}\right) \tag{4.13}$$

and

$$\left(\frac{\pi}{T}\right)^2 \geq \frac{1 - 3\xi_1}{1 - \xi_1} + \frac{\gamma}{2(1-\eta_1)^4}\left(3R_* - \frac{\gamma(1-\eta_1)^4}{2(1-\xi_1)^6}\right). \tag{4.14}$$

Below we present a numerical example of the results mentioned above.

Example 4.1 The following example is from [26]. Let us consider the differential equation

$$\ddot{x} + \frac{0.2}{(1-x)^3}\dot{x} + x = \frac{(0.25 + 0.1\cos(2\pi t))^2}{(1-x)^2}. \tag{4.15}$$

In this case $V(t) = 0.25 + 0.1\cos(2\pi t)$ is a positive 1-periodic function, such that $V_m = 0.15$ and $V_M = 0.35$. Furthermore, $0 < V_m^2 < V_M^2 \leq 4/27$. Then, by Theorem 4.3 we have the following:

1. Equation (4.15) admits a T-periodic solution $\psi_2(t)$ such that

$$\eta_2 \leq \psi_2(t) \leq \xi_2, \quad \forall t \in [0, 1],$$

where $\eta_2, \xi_2 \in [1/3, 1]$ are a pair of lower and upper solutions of (4.15) given as solution of the equations

$$\eta(1-\eta)^2 = 0.0225, \quad \text{and} \quad \xi(1-\xi)^2 = 0.1225, \qquad (*)$$

respectively. Hence, $\eta_2 \approx 0.5098$ and $\xi_2 \approx 0.8359$.

2. In order to prove the existence of a second 1-periodic solution lying in $[0, 1/3]$, notice that there exists a second pair of lower and upper $\eta_1, \xi_1 \in [0, 1/3]$ given as solutions of $(*)$ respectively. Hence, $\eta_1 \approx 0.1840$ and $\xi_1 \approx 0.0236$. Direct computation shows that

$$N = 0.3681, \quad \tilde{a} = 0.3681, \quad R_* = 0.3841, \quad \text{and} \quad M = 1.4715 \leq \pi^2.$$

From here, we are able to determine L_* as the unique solution of

$$\sin(\sqrt{L}) = \sqrt{L}\Big(\frac{L-M}{L+M}\Big),$$

therefore, $L_* = 3.964$. Finally

$$H(L_*) = \frac{L_* - M}{\sqrt{L_*}} \cot\Big(\frac{\sqrt{L_*}}{2}\Big) = 0.818, \quad \Rightarrow \quad N \leq H(L_*).$$

We have checked all the conditions of Theorem 4.3, in consequence, there exists a 1-periodic solution $\psi_1(t)$ of (4.15) such that

$$0.0236 \leq \psi_1(t) \leq 0.1840, \quad \forall t \in [0, 1].$$

Remark 4.6 If we assume that for the Nathanson graphene-based model the nonlinear damping forces are the same as in the canonical case, we obtain the differential equation

$$\ddot{x}(t) + \frac{\gamma}{(1-x(t))^3}\dot{x}(t) + x(t) - \alpha|x(t)|x(t) = \frac{eV^2(t)}{(1-x(t))^2}. \qquad (4.16)$$

Clearly, a combination of the hypothesis in Theorem 4.2 and Theorem 4.3 will provide the existence of at least three T-periodic solutions of (4.16), one negative and the other two positives. To our knowledge, up to today, there are no numerical or analytical studies of (4.16) on the multiplicity and stability of periodic solutions (Fig. 4.6).

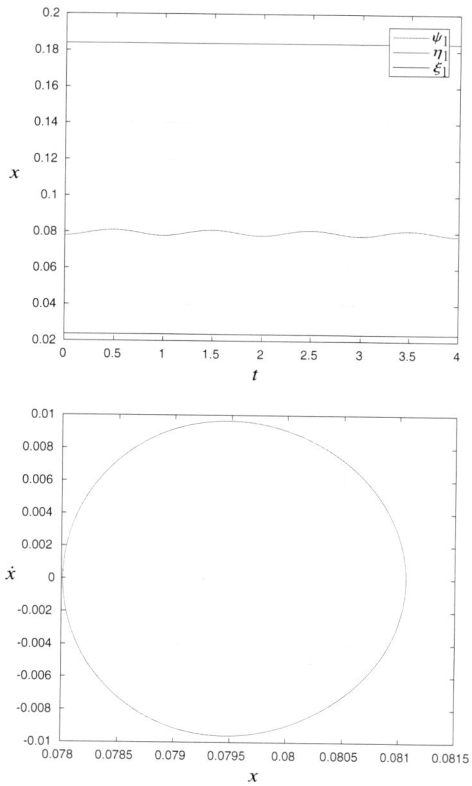

Fig. 4.6 Figure 4.1 in [26]. (Up): In black, the upper solution $\xi(t) = \xi_1$ for all $t \in \mathbb{R}$. In red, the lower solution $\eta(t) = \eta_1$ for all $t \in \mathbb{R}$. In blue, a numerical 1-periodic solution $\psi_1(t)$ of (4.15). (Down): Phase portrait of the numerical solution $\psi_1(t)$

4.3.3 Periodic Solution Under Delayed Controller Effects

As we establish in the introduction, in appropriate units, the equation of motion of the Nathanson model under a feedback time-delayed controller under a linear damping force and restoring force is given by

$$\ddot{x}(t) + \varrho\dot{x}(t) + x(t) = \frac{e[V(t) + \sigma(s(t), s_d(t), g)]^2}{(1 - x(t))^2}, \tag{4.17}$$

where

$$V(t) = V_0 + \delta v(t), \quad \text{and} \quad \sigma(s(t), s_d(t), g) = \begin{cases} g(x_d(t) - x(t)) & \text{if} \quad s(t) = x(t), \\ g(\dot{x}_d(t) - \dot{x}(t)) & \text{if} \quad s(t) = \dot{x}(t). \end{cases}$$

From now on let us assume that $s(t) = \dot{x}(t)$, i.e., the output signal that the time-delayed feedback controller acts over the velocity of the moveable part of the actuator. Throughout the method of multiples scales, the authors in [9, 10] study the dynamical response and stability of the actuator (4.17) for small values of δ and g. For convenience, we rewrite (4.17) in the form

$$\dot{X}(t) = F(t, X(t), X_d(t), g),$$

with

$$F(t, X, X_d, g) = \begin{pmatrix} \dot{x} \\ \frac{e[V(t)+\sigma(\dot{x},\dot{x}_d,g)]^2}{(1-x)^2} - x - \varrho\dot{x} \end{pmatrix}, \tag{4.18}$$

where $X(t) = (x(t), \dot{x}(t))^{tr}$ and $X_d(t) = (x_d(t), \dot{x}_d(t))^{tr}$. In the case, $d = 0$ or $g = 0$, the system (4.18) is reduced to the nonlinear system

$$\dot{X}(t) = F(t, X(t), 0), \quad \text{with} \quad F(t, X, 0) = \begin{pmatrix} \dot{x} \\ \frac{eV^2(t)}{(1-x)^2} - x - \varrho\dot{x} \end{pmatrix}. \tag{4.19}$$

From Theorem 4.2 (with $\alpha = 0$) and as well from Theorem 5 in [13], the system (4.19) admits exactly two T-periodic solutions

$$\Phi_1(t) = \begin{pmatrix} \varphi_1(t) \\ \dot{\varphi}_1(t) \end{pmatrix} \quad \text{and} \quad \Phi_2(t) = \begin{pmatrix} \varphi_2(t) \\ \dot{\varphi}_2(t) \end{pmatrix},$$

such that

$$0 < \xi_1 < \varphi_1(t) < \eta_1 \leq \frac{1}{3}, \quad \text{and} \quad \frac{1}{3} \leq \eta_2 < \varphi_2(t) < \xi_2 < 1, \quad \forall t \in \mathbb{R},$$

where

$$\xi_i(1-\xi_i)^2 = eV_m^2, \quad \text{and} \quad \eta_i(1-\eta_i)^2 = eV_M^2, \quad i = 1, 2.$$

Recently, the authors in [27] using the Implicit Function Theorem in Banach spaces proved the existence of periodic solutions of (4.17). The result establishes the following:

Theorem 4.4 (Theorem 9 in [27]) *Let $\Phi_i(t)$, $i \in \{1, 2\}$ be a T-periodic solution of (4.19). Then, there exists a neighborhood $\mathcal{V} \subset \mathbb{R}^2$ of $(0, 0)$ and a unique function $X^g(t) = X(t, d, g)$ that is a T-periodic solution of (4.18) for all $g \in \mathcal{V}$, i.e.,*

- ▷ $\dot{X}^g(t) = F(t, X^g(t), X_d^g(t), g)$, $\quad X^g(t+T) = X^g(t), \forall t \in \mathbb{R}, 0 < d < T$ *and* $G \in \mathcal{V}$.
- ▷ $X^0(t) = \Phi_i$ *for all* $t, d \in \mathbb{R}$, *with* $\Phi_i \in \{\Phi_1, \Phi_2\}$, *that is, for* $g = 0$.

Theorem 4.4 provides a local continuation of T-periodic solutions for (4.18) using g as the continuation parameter. This continuation method is also possible if instead of considering g as the continuation parameter, it is considered the delay $0 < d < T$.

In this scenario, it will be necessary some additional conditions that we found in [27].

Lemma 4.1 *Let $\Phi(t) = (\varphi(t), \dot{\varphi}(t))^{tr}$ be a T-periodic solution of the differential system* (4.19)*. Assume that $0 < \xi \leq \varphi(t) \leq \eta$ for all $t \in \mathbb{R}$. Then, $|\dot{\varphi}(t)| \leq \Lambda(\xi, \eta)$ for all $t \in \mathbb{R}$, where*

$$\Lambda(\xi, \eta) := \gamma(\eta - \xi) + \max\left\{\left|\frac{eV_m^2}{(1-\xi)^2} - \eta\right|, \left|\frac{eV_M^2}{(1-\eta)^2} - \xi\right|\right\} T. \tag{4.20}$$

Theorem 4.5 (Theorem 13 in [27]) *Let $\Phi_2(t) = (\phi_2(t), \dot{\phi}_2(t))^{tr}$ be the T-periodic solution of* (4.19)*.*[1] *For $g_2 > 0$, let us define*

$$b^* = \frac{2g_2(1-\xi_2)}{V_{\delta,\max}}, \quad and \quad \dot{b}^* = \frac{2eg_2}{\xi_2^3}\Big(\xi_2 \|\dot{V}_\delta\|_\infty + 2V_{\delta,\max}\Lambda(\xi_2, \eta_2)\Big),$$

with $\Lambda(\xi_2, \eta_2)$ given as in Lemma 4.1. Then, there exists a neighborhood $\Omega \subset \mathbb{R}$ of $d = 0$ and a unique function $X(t, d)$, that is a T-periodic solution of (4.18) *for all $d \in \Omega$, i.e.,*

- ▷ $\dot{X}(t, d) = F(t, X(t, d), X_d(t, d)), \quad X(t + T, d) = X(t, d)$, *for all $d \in \Omega$.*
- ▷ $X(t, 0) = \Phi_2(t)$ *for all $t \in \mathbb{R}$,*

if $g_2 > 0$ and one of the following conditions holds

(a) If $g_1 > 0$:

$$\begin{aligned} &\frac{1}{4}(c + b^*)^2 + \frac{1}{2}(\dot{b}^*)^2 \leq \frac{2g_1(1-\eta_2)}{V_{\delta,\min}} + \frac{3\xi_2 - 2}{\xi_2}, \\ &\frac{2g_1(1-\xi_2)}{V_{\delta,\max}} + \frac{3\eta_2 - 2}{\eta_2} < \left(\frac{\pi}{T}\right)^2 + \frac{c^2}{4}. \end{aligned} \tag{4.21}$$

(b) If $g_1 < 0$:

$$\begin{aligned} &\frac{1}{4}(c + b^*)^2 + \frac{1}{2}(\dot{b}^*)^2 \leq \frac{2g_1(1-\xi_2)}{V_{\delta,\max}} + \frac{3\xi_2 - 2}{\xi_2}, \\ &\frac{2g_1(1-\eta_2)}{V_{\delta,\min}} + \frac{3\eta_2 - 2}{\eta_2} < \left(\frac{\pi}{T}\right)^2 + \frac{c^2}{4}. \end{aligned} \tag{4.22}$$

Moreover, if $g_2 < 0$ and the damping coefficient c satisfies

[1] The conditions of existence of this T-periodic solution are given by Theorem 2 in [27].

$$c + \frac{2g_2(1-\eta_2)}{V_{\delta,\min}} < 0, \quad or \quad c + \frac{2g_2(1-\xi_2)}{V_{\delta,\max}} > 0, \tag{4.23}$$

and one of the following conditions hold:

(c) For $g_1 > 0$*:*

$$0 < g_2 + \frac{cV_{\delta,\max}}{1-\xi_2}, \qquad \frac{1}{4}(c+b^*)^2 + \frac{1}{2}(\dot{b}^*)^2 < \frac{2g_1(1-\eta_2)}{V_{\delta,\min}} + \frac{3\xi_2 - 2}{\xi_2} \quad and$$

$$\frac{2g_1(1-\xi_2)}{V_{\delta,\max}} + \frac{3\eta_2 - 2}{\eta_2} < \left(\frac{\pi}{T}\right)^2 + \frac{c^2}{4}.$$

(d) For $g_1 < 0$*:*

$$0 < g_2 + \frac{cV_{\delta,\max}}{1-\xi_2}, \qquad \frac{1}{4}(c+b^*)^2 + \frac{1}{2}(\dot{b}^*)^2 < \frac{2g_1(1-\xi_2)}{V_{\delta,\max}} + \frac{3\xi_2 - 2}{\xi_2} \quad and$$

$$\frac{2g_1(1-\eta_2)}{V_{\delta,\min}} + \frac{3\eta_2 - 2}{\eta_2} < \left(\frac{\pi}{T}\right)^2 + \frac{c^2}{4},$$

we obtained the same conclusion.

Remark 4.7 With the appropriate conditions, Theorem 4.5 can be extended to the periodic solution $\Phi_1(t)$, proving that the system (4.7) (equivalently Eq. (4.18)) admits exactly two positive T-periodic solution for $|g|$ small and $d \in]0, T[$ or for $d > 0$ small and g satisfying appropriate conditions.

4.3.4 Periodic Motions in the Comb-Drive Finger Oscillator

At this point, it is not a surprise to see that the previous methodology can be extended to solve mathematically the problem of the existence and stability of periodic solutions for the Comb-drive finger model (4.2). In fact, the authors in [11, 12, 17, 26] solve these questions for the linear damping case. Furthermore, in [17] consider a slight and very practical mechanical modification of (4.2) given by

$$\ddot{x}(t) + \varrho\dot{x}(t) + x(t) - u = \frac{4ex(t)V^2(t)}{(1-x^2(t))^2}, \tag{4.24}$$

where the parameter $u > 0$ represents a small deviation of the moveable finger from the middle line of the gap between the two fixed ones. These deviations emerge as a consequence of the micro-fabrication process of devices with small physical dimensions, for example within a range of 10^{-7}m to 10^{-3}m. Notice that $u = 0$, if $x(t)$ is a solution of (4.24) then $-x(t)$ is also a solution. Moreover, in this case, the Comb-drive finger oscillator admits the trivial solution $x(t) = 0$ for all $t \in \mathbb{R}$.

For a fixed $u \in [0, 1[$ let us define the following quantities

$$\vartheta_m := 4eV_m^2, \quad \vartheta_M := 4eV_M^2, \quad \text{and} \quad \vartheta_*(u) = (1-\varsigma(u)^2)(5\,\varsigma(u)^2-4\,u\,\varsigma(u)-1),$$

with $\varsigma(u)$ and $\psi(u)$ given by

$$\varsigma(u) = \frac{u^{1/3}}{4}\left(u^{2/3} + \psi(u) + \frac{u^{4/3}}{\psi(u)}\right) \quad \text{and} \quad \psi(u) = \left(u^2 + 4u\,\sqrt{u^2+4} + 8\right)^{1/3}. \tag{4.25}$$

Lemma 4.2 *Consider the Eq. (4.24) with $u \in [0, 1[$ fixed. Assume that*

$$0 < \vartheta_m \leq \vartheta_M \leq \vartheta_*(u),$$

with $\vartheta_*(u)$ given by (4.25) and let $\phi(x, u) = (x-u)(1-x^2)^2$. Then, there exist three constant upper solutions δ_i, $i = 0, 1, 2$ satisfying

$$\phi(\delta_0, u) = \vartheta_{\max}\delta_0, \quad \phi(\delta_i, u) = \vartheta_{\min}\delta_i, \quad i = 1, 2,$$

and three constant lower solutions ζ_j, $j = 0, 1, 2$ satisfying

$$\phi(\zeta_0, u) = \vartheta_{\min}\zeta_0, \qquad \phi(\zeta_i, u) = \vartheta_{\max}\zeta_i, \quad i = 1, 2.$$

Moreover,

$$-1 < \zeta_0 \leq \delta_0 < 0, \quad \text{and} \quad 0 < \delta_1 \leq \zeta_1 < \varsigma(u) < \zeta_2 \leq \delta_2 < 1.$$

Theorem 4.6 *Let us consider $u \in [0, 1[$ fixed. Under the conditions of Lemma 4.2, and in addition with*

$$M := 1 - \vartheta_m < \left(\frac{\pi}{2}\right)^2 \quad \text{and} \quad \varrho \leq \varrho^* := \max_{L\in[M,\pi^2/T^2]} \frac{L-M}{\sqrt{L}}\cot\left(\frac{\sqrt{L}T}{2}\right). \tag{4.26}$$

There exist at least three non-trivial T-periodic solutions $\rho(t)$, $\psi_1(t)$ y $\psi_2(t)$ of (4.24) such that

$$\xi_0 \leq \rho(t) \leq \delta_0, \qquad \delta_1 \leq \psi_1(t) \leq \xi_1 \quad \text{and} \quad \xi_2 \leq \psi_2(t) \leq \delta_2,$$

for all $t \in [0, T]$. Furthermore:

- *If $u = 0$ then $\psi_1(t) = 0$ and $\rho(t) = -\psi_2(t)$ for all $t \in \mathbb{R}$.*
- *For $u \ll 1$, the only non-trivial positive T-periodic solution of (4.24) are $\psi_1(t)$ and $\psi_2(t)$, meanwhile, the only non-trivial negative T-periodic solution is $\rho(t)$.*

- *If $\varrho \le \varrho^*$ and $u \ll 1$, then $\rho(t)$ and ψ_2 are unstable, meanwhile $\psi_1(t)$ is locally asymptotically stable if $\varrho \le \min\{\varrho^*, \varrho^{**}\}$ with*

$$\varrho^{**} = 2\sqrt{1 - \vartheta_{\max} s'(\xi_1)}, \quad \text{where} \quad s(x) = \frac{x}{(1-x^2)^2}.$$

Example 4.2 Taken reference values from [4], the authors in [17] consider a shifted Comb-drive oscillator with the following parameters

$$\varrho = \frac{6}{1000}, \quad e = \frac{5}{256} \quad u = 0.1, \quad \text{and} \quad V(t) = \frac{3}{2} + \frac{1}{2}\cos\left(\frac{4\pi}{3}t\right).$$

With these choice of parameters, it can be verified that

$$\varrho^* \approx 0.0068, \quad \sigma^{**} \approx 0.6438, \quad \text{and} \quad \varrho \le \min\left\{\varrho^*, \varrho^{**}\right\}.$$

Notice that $V(t)$ is a positive $T = \frac{3}{2}$-periodic function, with $V_m = 1 \le V(t) \le 2 = V_M$ for all $t \in \mathbb{R}$. The search for upper and lower solutions in $[0, 1]$ by seeking solutions of the equations

$$\varphi(x, 0.1) - \frac{20}{256}x = 0, \quad \text{and} \quad \varphi(x, 0.1) - \frac{80}{256}x = 0, \tag{4.27}$$

respectively. By Theorem 4.6 it can be verified that for the corresponding shifted Comb-drive oscillator, there are three constant upper solutions in $[-1, 1]$, namely,

$$\delta_0 = -0.69102212..., \quad \delta_1 = 0.10869616..., \quad \text{and} \quad \delta_2 = 0.83794673....$$

Similarly, there are three constant lower solutions in $[-1, 1]$. These are

$$\zeta_0 = -0.85760592..., \quad \zeta_1 = 0.14853528..., \quad \text{and} \quad \zeta_2 = 0.62451360...$$

From Theorem 4.6, the existence of three $3/2$-periodic solutions, ρ, ψ_1, and ψ_2 that satisfy

$$\zeta_0 \le \rho(t) \le \delta_0, \qquad \delta_1 \le \psi_1(t) \le \zeta_1, \qquad \zeta_2 \le \psi_2(t) \le \delta_2,$$

is guaranteed. Looking for $3/2$-periodic solutions that satisfy $x(0) = x_0$, $\dot{x}(0) = 0$, and $\dot{x}(3/2) = 0$, the numerical approximation in [17] provides the following

$$\rho(0) = -0.7542859, \quad \dot{\rho}(0) = 0, \quad \text{and} \quad |\rho(0) - \rho(1.5)| \approx 3 \times 10^{-4},$$

$$\psi_1(0) = 0.12268178, \quad \dot{\psi}_1(0) = 0, \quad \text{and} \quad |\psi_1(0) - \psi_1(1.5)| \approx 1 \times 10^{-6},$$

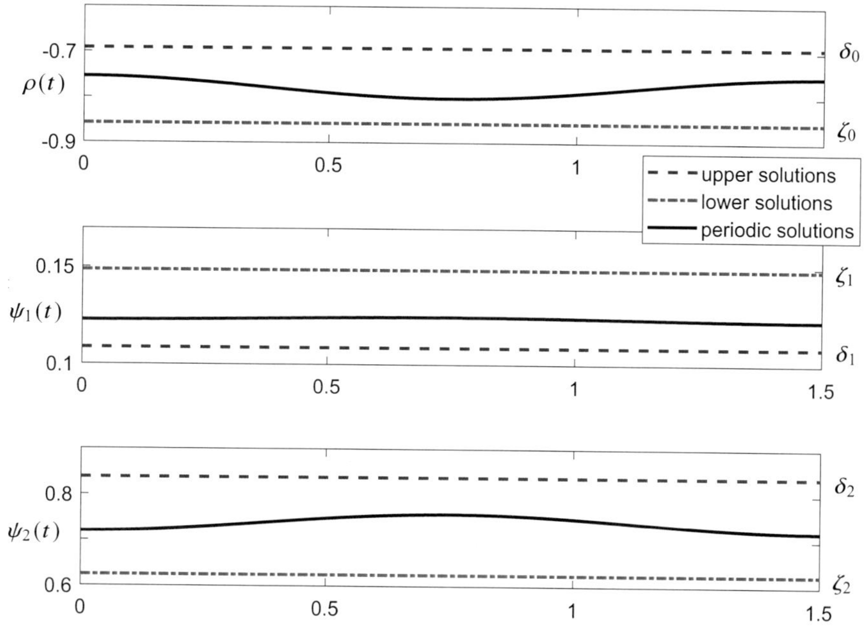

Fig. 4.7 Figure 9 in [17]. Numerical approximation of the three 3/2-periodic solutions $\rho(t)$, $\psi_1(t)$ and $\psi_2(t)$ of the Comb-drive oscillator (4.24) with parameters $\varrho = 0.006$, $e = 5/256$ and voltage source $V(t) = 0.5\big(3 + \cos\big(4\pi t/3\big)\big)$ given by Theorem 4.6

and finally

$$\psi_2(0) = 0.7195052, \quad \dot{\psi}_2(0) = 0, \quad \text{and} \quad |\psi_2(0) - \psi_2(1.5)| \approx 1.7 \times 10^{-4}.$$

The numerical approach of these functions is plotted in Fig. 4.7 along with their corresponding upper and lower solutions.

Let us consider $V(t, \delta) = 0.5\big(3 + \delta \cos\big(4\pi t/3\big)\big)$, a parametric voltage source, with $|\delta| < 3$. It is possible to find a family of 3/2-periodic solutions of (4.24) with initial conditions $x(0) = x_0$, $\dot{x}(0) = 0$ and $\dot{x}(T) = 0$ for many values of δ. The results are in the upper left graph in Fig. 4.8, in which each point on the graph represents a pair (x_0, δ) that results in a 3/2-periodic solution of (4.24).

For solutions with initial conditions $x(0)$ in $[0, 1]$, then $\delta \in [P_1, P_2]$ with $P_1 \approx -6.4683$ and $P_2 \approx 6.4657$. Concerning the linear stability properties of these periodic solutions, in [17] analyze the discriminant Δ of the linearized system, throughout the classical Floquet theory (see [28]). Recall that if $|\Delta| < 2$ the corresponding periodic solutions will be linearly stable and unstable if $\Delta > 2$. The other three plots in Fig. 4.8 show the results of this computation. The branch in blue (lower left) corresponds to stable solutions $(0 < \Delta < 2)$ and the branches in red (upper and lower right) correspond to unstable solutions $(\Delta > 2)$.

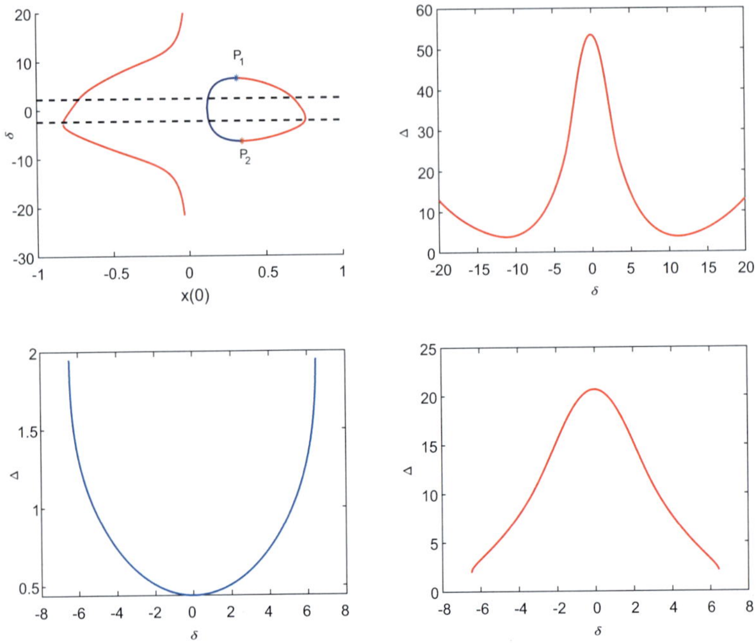

Fig. 4.8 Figure 10 in [17]. Upper left: Family of 3/2-periodic solutions. The dashed lines represent the range of δ values that apply to Theorem 4.6. Upper right: Plot of δ vs Δ for negative initial conditions. Lower Left (Right): Plot of δ vs Δ for positive(negative) initial conditions that result in stable solutions

4.4 Periodic Motions with Non-constant Sign

In Sect. 4.3, throughout the method of upper and lower solutions, we showed that the actuators (4.1) and (4.2) admit periodic solutions. These particular solutions share the same qualitative property: *All have a defined sign*. In this section, we will show two results about periodic oscillations of (4.1) and (4.2) using a different approach. Let us begin with the Nathanson actuator with the equation of motion given by

$$\ddot{x}(t) + x(t) + \varepsilon\Big(\dot{x}(t) - \frac{eV^2(t)}{(1 - x(t))^2}\Big) = 0, \tag{4.28}$$

with ϵ positive and small. Equation (4.28) comes from (4.7), under the assumption of a low viscosity damping and low voltage regime, as follows

$$\varrho \to \boldsymbol{e}, \quad \text{and} \quad V(t) \to \varepsilon V(t).$$

Now, suppose that $V(t)$ is a $\dfrac{2p\pi}{q}$-periodic function with q, p positive integers and co-prime. Using the Averaging Theory (see [29]) the authors in [11] proved the existence of periodic solutions of (4.28) with a non-constant sign as bifurcations of periodic solutions of the harmonic oscillator

$$\ddot{x}(t) + x(t) = 0. \tag{4.29}$$

Their result is as follows:

Theorem 4.7 *Let $V = V(t) \in C(\mathbb{R}/T\mathbb{Z})$ with $T = \dfrac{2p\pi}{q}$ where q and p are co-prime positive integers. Define the functions*

$$\Upsilon(t, x_0, y_0) = \frac{1}{2p\pi}\left(\dot{x}_0 \cos t - x_0 \sin t - \frac{eV^2(t)}{(1 - x_0 \cos t - y_0 \sin t)^2}\right),$$

and

$$f(x_0, y_0) = -\int_0^{2p\pi} \Upsilon(t, x_0, y_0) \sin t\, dt, \qquad g(x_0, y_0) = \int_0^{2p\pi} \Upsilon(t, x_0, y_0) \cos t\, dt.$$

Then, for every $\varepsilon \neq 0$ sufficiently small, and for every real solution $(\hat{x}_0, \hat{y}_0)$ of the system

$$f(x_0, y_0) = 0 \quad \textit{and} \quad g(x_0, y_0) = 0, \tag{4.30}$$

satisfying

$$\det \mathcal{H} \neq 0, \quad \textit{with} \quad \mathcal{H} = \begin{pmatrix} \partial_{x_0} f(\hat{x}_0, \hat{y}_0) & \partial_{y_0} f(\hat{x}_0, \hat{y}_0) \\ \partial_{x_0} g(\hat{x}_0, \hat{y}_0) & \partial_{y_0} g(\hat{x}_0, \hat{y}_0) \end{pmatrix},$$

the Nathanson oscillator (4.28) *has a $2p\pi$-periodic solution $x(t, \varepsilon)$ such that, if $\varepsilon \to 0$ then $x(t, \varepsilon) \to x(t, \hat{x}_0, \hat{y}_0)$ where*

$$x(t, \hat{x}_0, \hat{y}_0) = \hat{x}_0 \cos t + \hat{y}_0 \sin t, \quad \hat{x}_0, \hat{y}_0 \in \mathbb{R},$$

is the solution of the harmonic oscillator (4.29) *with initial conditions $x(0) = \hat{x}_0$ and $\dot{x}(0) = \hat{y}_0$ and traveled p-times around the origin. Moreover, if the real part of all the eigenvalues of the matrix $\mathcal{H}$ are negative, the periodic solution $x(t, \varepsilon)$ is stable, meanwhile, if some of those real parts are positive, the periodic solution $x(t, \varepsilon)$ is unstable.*

Remark 4.8 It is worth mentioning that Theorem 7 does not require condition $C1$ over $V(t)$, but the result is developed for a small amplitude voltage load. Corollary 1 in [11] provides a numerical example of a unique $T = 2\pi$-periodic solution ($p = q = 1$) of (4.28) given in Theorem 4.7 with voltage source $V(t) = \dfrac{1}{6\sqrt{3}}\big(2 + \cos t\big)$. This periodic solution is associated with the unique zero

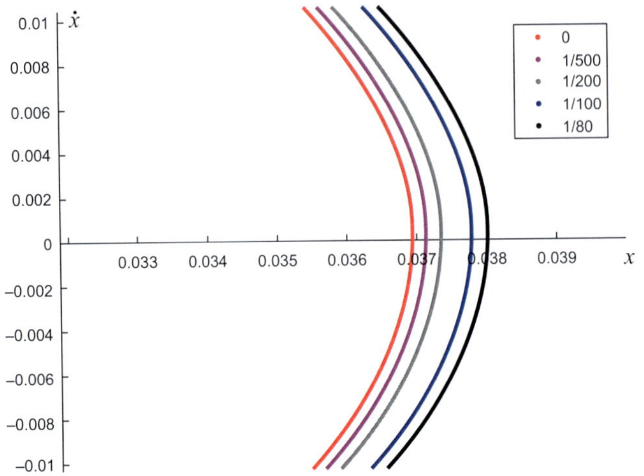

Fig. 4.9 Figure 3 in [11]. In red, graph of the solution $\hat{x}(t) = \hat{x}_0 \cos t + \hat{y}_0 \sin t$. of (4.29) in the phase plane $x, \dot{x}$ restricted to the box $[0.032, 0.039] \times [-0.01, 0.01]$. The other four orbits are the graph of the 2π-periodic solutions $x(t, \varepsilon_i)$ of (4.28) with initial conditions $(\hat{x}_0, \hat{y}_0) + O(\varepsilon_i)$ with $\varepsilon_1 = 1/80$ (black), $\varepsilon_2 = 1/100$ (blue), $\varepsilon_3 = 1/200$ (gray) and $\varepsilon_4 = 1/500$ (magenta) close to $\hat{x}(t)$

of the (4.30) given numerically by

$$P = (\hat{x}_0, \hat{y}_0) = (-0.0368\ldots, -0.0028\ldots).$$

Since $\det \mathcal{H} < 0$ at P, then the corresponding 2π-periodic solutions is unstable. In Fig. 4.9 plotted four different numerical solutions of (4.28), namely, $x(t, \varepsilon_i)$, $i = 1, 2, 3, 4$, with initial conditions very close to P. Notice how for $\varepsilon \to 0$ these solutions tend to $\hat{x}(t) = \hat{x}_0 \cos t + \hat{y}_0 \sin t$. Finally, it can be proved that for the non-shifted Comb-drive finger model. namely, the model (4.24) with $u = 0$, the trivial solution $x \equiv 0$ is locally asymptotically stable, making impossible the existence of periodic orbits near the origin. This is why Averaging Theory applied to the non-shifted Comb-drive model does not provide information about periodic orbits.

4.4.1 Comb-Drive Oscillator with Cubic Stiffness

We finish this section by considering a Comb-drive finger model (4.2) with the following assumptions

$$c(x) = 0, \quad \text{and} \quad q(x) = x + \alpha x^3, \quad \alpha \in \mathbb{R}^+.$$

In such a case, the equation of motion of the actuator is given by

$$\ddot{x}(t) + x(t) + \alpha x^3(t) = \frac{4eV^2(t)}{(1 - x^2(t))^2}, \tag{4.31}$$

with $|x(t)| < 1$ for all $t \in \mathbb{R}$. Recall that $V(t)$ is a positive T-periodic function. Here we assume the following additional condition:

C3. $V(t)$ *is an even function* $\Leftrightarrow$ $V(t) = V(-t), \quad \forall t \in \mathbb{R}.$

For the comb-drive finger oscillator (4.31) under the conditions $C1, C2$ (see Sect. 4.3) and $C3$, the authors in [30] prove the existence of odd periodic solutions with a prescribed number of oscillations. The basic tool for achieving these results is the recent method proposed by R. Ortega in [31] concerning the existence of solutions to the Dirichlet problem

$$\ddot{x}(t) + x(t)D(t, x(t)) = 0, \quad x(0) = x(L) = 0, \tag{4.32}$$

with $D = D(t, x) \in C^{0,1}([0, L] \times \mathbb{R})$ for some $L > 0$, where $C^{0,1}([0, L] \times \mathbb{R})$ denotes the class of functions such that $\partial_x D(t, x)$ exists everywhere and $(t, x) \to D(t, x)$ and $(t, x) \to \partial_x D(t, x)$ are continuous on $[0, L] \times \mathbb{R}$. Moreover,

- ▷ $D(t, x) < D(t, 0), \quad x \neq 0,$
- ▷ $|xD(t, x)| \leq C, \quad (t, x) \in [0, L] \times \mathbb{R},$

where $C > 0$ is a fixed constant. Ortega's idea also considers the variational equation associated with (4.32) at $x = 0$, given by

$$\ddot{z} + D(t, 0)z = 0. \tag{4.33}$$

In addition, for the solution $\tilde{z}(t)$ of (4.33) that satisfies the initial conditions $z(0) = 0$, $\dot{z}(0) = 1$ define the number ν_0 as the cardinality of the set

$$Z = \{t \in]0, L[: \tilde{z}(t) = 0\},$$

which if well-defined since $\tilde{z}(t)$ have isolated zeros by uniqueness. This method is presented in the Appendix as *Ortega's Principle*, named in this way by the authors in [30].

Let us see how the problem of the existence of periodic solutions of (4.31) was wisely adapted by authors in [30] to the existence of solutions of a Dirichlet problem like (4.32).

An Associated Dirichlet Problem The first step is to rewrite (4.31) as follows

$$\ddot{x}(t) + x(t)\mathcal{D}(t, x(t)) = 0, \tag{4.34}$$

where $\mathcal{D} \in C^{0,1}(\mathbb{R}\times] - 1, 1[)$, $(t, x) \to \mathcal{D}(t, x)$ where

$$\mathcal{D}(t,x) = 1 + \alpha x^2 - \frac{4eV^2(t)}{(1-x^2)^2}.$$

Observe that the function $\mathcal{D}$ verifies the following symmetries

$$\mathcal{D}(-t,x) = \mathcal{D}(t,x), \quad \text{and} \quad \mathcal{D}(t,x) = \mathcal{D}(t,-x),$$

for all $(t,x) \in \mathbb{R}\times]-1,1[$. Given a positive integer p, it is well know that the symmetries allow to confirm that the search of non-trivial odd pT-periodic solutions of (4.34) can be reduced to the existence of non-trivial solutions of the Dirichlet problem

$$\ddot{x}(t) + x(t)\mathcal{D}(t,x(t)) = 0, \quad x(0) = x(pT/2) = 0. \tag{4.35}$$

Bounds for Any Periodic Solution The second step is to find *a priori bound for any periodic solution* $\varphi(t)$ *of* (4.34). Therefore, in [30] we found

$$|\varphi(t)| < r_0 + \epsilon_0, \quad \forall t \in \mathbb{R},$$

with r_0 the unique solution in $]0,1[$ of the equation

$$(1+\alpha x^2)(1-x^2)^2 = 4eV_m^2,$$

and $\epsilon_0 \in \mathbb{R}^+$, such that $\epsilon_0 < 1 - r_0$. This bound is deduced from Lemma 4.3.

Lemma 4.3 (Lemma 3.1 in [30]) *Let us consider* $F : \mathbb{R}\times]a,b[\to \mathbb{R}$ *a continuous and Locally Lipschitz function, and the differential equation*

$$\ddot{x} = F(t,x). \tag{4.36}$$

Assume that F *satisfies the following*

a. F *is* T*-periodic in* t*, i.e.,* $F(t+T,x) = F(t,x)$ *for all* $(t,x) \in \mathbb{R}\times]a,b[$.
b. There exist $R_1, R_2 \in]a,b[$, $R_1 < R_2$ *such that for all* $t \in \mathbb{R}$:

$$F(t,x) < 0, \quad \forall x \in]a, R_1[, \quad \text{and} \quad F(t,x) > 0, \quad \forall x \in]R_2, b[$$

Then, any T*-periodic solution* $\varphi(t)$ *of* (4.36) *satisfies*

$$R_1 \le \varphi(t) \le R_2,$$

for all $t \in \mathbb{R}$.

It can be said that Lemma 4.3 is inspired by the well-known Hartman conditions (see [32]). An analogous result of a priori bound for periodic solutions of (4.31) with $\alpha = 0$ can be found in Proposition 7 in [33].

An Equivalent Equation The third and last step is to introduce a differential equation like in (4.32) well-defined for all $(t, x) \in \mathbb{R} \times \mathbb{R}$ and equivalent to (4.34) for $|x| < 1$. The idea is inspired by the truncation techniques for boundary value problems given in [32]. In [30] define the odd increasing function $\mathcal{T} \in C^1(\mathbb{R})$ given by

$$\mathcal{T}(x) = \begin{cases} \mathcal{T}_\mu(x), & \text{if} \quad x \geq 0, \\ -\mathcal{T}_\mu(-x), & \text{if} \quad x < 0, \end{cases}$$

where

$$\mathcal{T}_\mu(x) = \begin{cases} x & \text{if} \quad 0 \leq x \leq R_0, \\ \frac{\mu R_0(x - R_0) + x}{1 + \mu(x - R_0)}, & \text{if} \quad x > R_0. \end{cases}$$

The key properties of $\mathcal{T}(x)$ is that: $|\mathcal{T}(x)| < 1$ for all $x \in \mathbb{R}$ and $\mathcal{T}(x) = x$ for all $|x| \leq R_0$. Thus, the equivalent truncated equation for (4.34) is

$$\ddot{x}(t) + x(t)\hat{\mathcal{D}}(t, x(t)) = 0, \quad \text{for} \quad \hat{\mathcal{D}}(t, x) = \begin{cases} \frac{\mathcal{T}(x)}{x}\mathcal{D}(t, x), & \text{if} \quad x \neq 0, \\ \mathcal{D}(t, 0), & \text{if} \quad x = 0. \end{cases} \tag{4.37}$$

It can be proved that $\mathcal{D} \in C^{0,1}(\mathbb{R} \times \mathbb{R})$ and satisfies the hypothesis of Ortega's principle. Now we are able to present the main result in [30].

Theorem 4.8 (Theorem 2.2 in [30]) *Consider $p > 0$ and $q \geq 0$ two integers numbers, and assume that the following conditions hold*

$$i) \quad 4eV_M^2 < 1, \quad \text{and} \quad ii) \quad \alpha \leq 8eV_m^2.$$

Then, the Eq. (4.31) has an odd pT-periodic solution $\varphi_q(t)$ with q zeros in $]0, pT/2[$ such that $\varphi_q(0) > 0$ if and only if $\nu_0 > q$ where ν_0 is the number of zeros of $\tilde{z}$ in $]0, pT/2[$, the solution of the initial value problem

$$\ddot{z} + (1 - 4eV^2(t))z = 0, \quad z(0) = 0, \quad \dot{z}(0) = 1.$$

Moreover, if $\nu_0 > 0$ then $\varphi_0(t)$ is the unique odd pT-periodic solution of (4.31) positive in $]0, pT/2[$ and therefore has minimal period pT.

Corollary 4.1 (Corollary 2.3 in [30]) *Consider p and q_0 two positive integers numbers, and assume that the conditions of Theorem 4.8 hold. Assume that $T > \dfrac{2\pi(q_0 + 1)}{p}$ and*

$$0 < 1 - \left(\frac{2\pi(q_0+1)}{pT}\right)^2 \leq 4eV_m^2 < 4eV_M^2 \leq 1 - \left(\frac{2\pi q_0}{pT}\right)^2.$$

Then, $\nu_0 = q_0$ *or* $\nu_0 = q_0 + 1$, *and* (4.31) *has and odd* pT*-periodic solution* φ_q *with* q *zeros in* $]0, pT/2[$ *for each* $q = 0, 1, 2, \ldots, q_0 - 1$.

4.5 The Mathematical Toolbox

To provide a self-contained document, in this section, we briefly establish some of the fundamental results of the theoretical framework for the main results given in Sect. 4.3. Here the reader can find the key points behind the proofs and the methodology used in [11–13, 15, 16].

4.5.1 *Existence of Periodic Solutions: The Method of Upper and Lower Solutions*

Consider the boundary value problem

$$\ddot{x} + F(t, x, \dot{x}) = 0, \quad x(0) = x(T), \quad \dot{x}(0) = \dot{x}(T) = 0, \tag{4.38}$$

where $F : \mathbb{R} \times \Omega \to \mathbb{R}$, is a continuous function and $\Omega \subseteq]x_1, x_2[\times\mathbb{R}$ is an open connected set with $-\infty \leq x_1 < x_2 \leq \infty$.

Definition 4.1 (Upper and Lower Solution) A function $\xi \in C^2(]0, T[) \cap C^1([0, T])$ is called upper solution of (4.38) relative to the domain $\mathbb{R} \times \Omega$ if

1. $\ddot{\xi}(t) + F(t, \xi(t), \dot{\xi}(t)) \leq 0$ for all $t \in]0, T[$.
2. $\xi(t) \in]x_1, x_2[$ for all $t \in]0, T[$, and $\xi(0) = \xi(T)$, $\dot{\xi}(0) \leq \dot{\xi}(T)$.

A function $\eta \in C^2(]0, T[) \cap C^1([0, T])$ is called a lower solution of (4.38) relative to the domain Ω if all the previous conditions hold with the reverse inequalities.

The upper and lower solutions are *well-ordered* if

$$\eta(t) \leq \xi(t), \quad \forall t \in [0, T].$$

Meanwhile, the upper and lower solutions are in the *reversed order* if

$$\xi(t) \leq \eta(t), \quad \forall t \in [0, T].$$

The search for upper and lower solutions of any differential equation is, in general, a difficult problem. For example, we can appreciate in [34, 35] the effort involved in the construction of non-trivial upper and lower solutions for Liénard type equations with singularities. It may be stated that the simplest upper and lower solutions are constant functions, i.e., $\xi(t) = \xi$ for all $t \in \mathbb{R}$ and $\eta(t) = \eta$ for all $t \in \mathbb{R}$. In such a case, the following inequalities hold:

$$F(t, \xi, 0) \leq 0, \quad \text{and} \quad 0 \leq F(t, \eta, 0), \quad \forall t \in [0, T]. \tag{4.39}$$

Therefore, if for all $t \in \mathbb{R}$ we have

$$F_*(x, \dot{x}) \leq F(t, x, \dot{x}) \leq F^*(x, \dot{x}), \quad \forall (x, \dot{x}) \in \Omega,$$

the inequalities in (4.39) hold if $F^*(\xi, 0) \leq 0$ and $0 \leq F_*(\eta, 0)$. In particular, a constant upper solution and a constant lower solution can be selected as the solution to the equations

$$F_*(\xi, 0) = 0, \quad \text{and} \quad F^*(\eta, 0) = 0,$$

respectively. For example, if we apply this technique to the case

$$F(t, x, \dot{x}) = c(t, x)\dot{x} + q(x) - \frac{G(t)}{K(x)}, \quad \text{with} \quad (x, \dot{x}) \in \Omega,$$

the upper and lower solutions can be computed as the solutions of

$$q(\xi)K(\xi) - G_m = 0, \quad \text{and} \quad q(\eta)K(\eta) - G_M = 0.$$

This is precisely the fundamental idea for the existence of constant upper and lower solutions of the Eqs. (4.7), (4.10), and (4.12). Before showing some lines of the classical method of upper and lower solutions, the following theorem due to M. Nagumo [36] provides an interesting result about the bound for the derivative of smooth functions.

Theorem 4.9 *Let $\rho : \mathbb{R}_+ \to \mathbb{R}_+ \setminus \{0\}$ be a continuous function such that*

$$\int_0^\infty \frac{s}{\rho(s)} ds = \infty.$$

Assume that $x \in C^2([0, T], \mathbb{R})$ is such that

$$|x(t)| \leq K, \quad \textit{and} \quad |\ddot{x}(t)| \leq \rho(|\dot{x}(t)|), \quad \forall t \in [0, T]$$

for some $K > 0$. Then, there exists a number $\hat{R} = \hat{R}(T, K, \rho)$, such that $\|\dot{x}\|_\infty \leq \hat{R}$.

Theorem 4.9 provides *a priori* bound of the derivative of $x(t)$ by assuming some appropriate bounds of $|x(t)|$ and $|\ddot{x}(t)|$. This result can be extended to a second-order differential equation in the following way.

Given $a, b \in C([0,T])$ such that $a(t) \leq b(t)$ for all $t \in [0,T]$, define the set

$$E_{a,b} := \Big\{(t,x,y) \in [0,T] \times \mathbb{R}^2 \mid a(t) \leq x \leq b(t)\Big\}.$$

Definition 4.2 (Nagumo Condition) Let $F : E_{a,b} \to \mathbb{R}$ continuous. The function F satisfies the Nagumo condition on $E_{a,b}$ if there exists a positive continuous function $\rho : \mathbb{R}^+ \to \mathbb{R}^+$ such that

$$\int_0^\infty \frac{s}{\rho(s)} ds = \infty \quad \text{and} \quad |F(t,x,y)| \leq \rho(|y|),$$

for all $(t,x,y) \in E_{a,b}$.

Application to Second-Order Differential Equations Let us see how the results presented in this section apply to different nonlinear oscillators, in particular to those presented in Sect. 4.2. Firstly, let us define

$$a_m := \min_{[0,T]} a(t), \quad \text{and} \quad b_M := \max_{[0,T]} b(t).$$

Let $F : E_{a,b} \to \mathbb{R}$, $F(t,x,y) = c(t,x)y + q(t,x)$ with $c, q \in C(\Delta)$ where $\Delta = [0,T] \times [\eta_m, \xi_M]$. Define $\tilde{\alpha} := \max\left\{\max_{E_{a,b}} |c(t,x)|, \max_{E_{a,b}} |q(t,x)|\right\}$ and the positive continuous function $\rho(s) = \tilde{\alpha}(s+1)$, $s \in \mathbb{R}^+$. Notice that

$$\int_0^\infty \frac{s}{\rho(s)} ds = \int_0^\infty \frac{s}{\tilde{\alpha}(s+1)} ds = \infty,$$

and

$$|F(t,x,y)| \leq \tilde{\alpha}(|y|+1) = \rho(|y|), \quad \forall (t,x,y) \in E_{a,b},$$

which implies that $F(t,x,y)$ satisfies a Nagumo condition on $E_{a,b}$. In consequence any Liénard differential equation of type

$$\ddot{x} + c(t,x)\dot{x} + q(t,x) = 0$$

satisfies a Nagumo condition on $E_{a,b}$.

There are several versions of the Nagumo condition. In [26], we found a practical and more useful Nagumo-type condition that dictates the following: *Let $F : E_{a,b} \to \mathbb{R}$ continuous. Then F satisfies a Nagumo condition on $E_{a,b}$ if for some continuous function $\rho : [0,\infty] \to [0,\infty]$ we have*

$$(\mathcal{N}) \quad \int_0^{\hat{R}} \frac{s}{\rho(s)} ds > 2\max\{\|a\|_\infty, \|b\|_\infty\} \quad \text{and} \quad |F(t,x,y)| \leq \rho(|y|),$$

for some $\hat{R} > 0$ and for all $(t,x,y) \in E_{a,b}$.

Next, we present the main theorems that give us sufficient conditions for the existence of solutions to the boundary value problem (4.38). Recall that, if F is T-periodic in t, every solution of (4.38) will be a T-periodic function. The proof of both theorems can be found in [32, 37].

Theorem 4.10 (Existence Under Direct Order) *Let η and ξ be lower and upper solution of* (4.38) *such that $\eta(t) \leq \xi(t)$ for all t in* $[0, T]$. *If F satisfies the Nagumo-type condition ($\mathcal{N}$) in $E_{\eta,\xi}$. Then, the boundary value problem* (4.38) *has at least one solution $\varphi \in C^2([0, T])$ such that*

$$\|\dot{\varphi}\|_\infty < \hat{R}_*, \quad \eta(t) \leq \varphi(t) \leq \xi(t), \qquad \forall t \in [0, T],$$

where $\hat{R}_* > \max\left\{\hat{R}, \|\eta\|_\infty, \|\xi\|_\infty\right\}$.

Theorem 4.11 (Existence Under Reversed Order) *Let $\eta, \xi \in C^2([0, T])$ be a lower and an upper solution of the boundary value problem* (4.38) *such that $\xi(t) \leq \eta(t)$ for all t in* $[0, T]$. *Define*

$$E_{\xi,\eta} := \left\{(t, x, y) \in \Omega \mid t \in [0, T],\ \xi(t) \leq x \leq \eta(t)\right\}.$$

(†) *Assume that there exists $N \geq 0$ such that $|\partial_y f(t, x, y)| \leq N$ for all (t, x, y) in $E_{\xi,\eta}$. Then for any solution $u(t)$ of* (4.38) *such that $\xi(t) \leq u(t) \leq \eta(t)$ on* $[0, T]$, *there exist $R := R(N) > 0$ such that $-R \leq \dot{u}(t) \leq R$ for all t in* $[0, T]$.

(††) *If the assumption in (†) holds. Let*

$$W := \left\{(t, x, y) \in E_{\xi,\eta} \mid t \in [0, T],\ -R \leq y \leq R\right\}, \tag{4.40}$$

and assume that there exists $M \geq 0$ such that for all (t, x, y) in W

$$\partial_x f(t, x, y) \leq M \leq \left(\frac{\pi}{T}\right)^2, \quad \sigma(t) \leq L \quad \text{and} \quad N \leq H(L),$$

for some $L \in \left[M, (\pi/T)^2\right]$ where

$$\sigma(t) = \frac{f(t, \eta(t), \dot{\eta}(t)) - f(t, \xi(t), \dot{\xi}(t))}{\eta(t) - \xi(t)} \quad \text{and} \quad H(L) = \frac{(L - M)}{\sqrt{L}} \cot\left(\frac{T\sqrt{L}}{2}\right).$$

Then the boundary value problem (4.38) *has at least one solution ψ such that*

$$\xi(t) \leq \psi(t) \leq \eta(t), \quad \forall t \in [0, T].$$

Remark 4.9 Theorem 4.10 can be applied to Eqs. (4.7), (4.10), and (4.12) if for each model there exist a pair of upper and lower solutions in direct order. This is why, the condition on the maximum voltage $V_M < V_{pull}$ is imposed (see Remark 4.5) in order to have a pair of constant upper and lower $\hat{\xi}$ and $\hat{\eta}$ respectively, and such that $\hat{\eta} < \hat{\xi}$. In consequence, the existence of at least one T-periodic solution of the Nathanson models (4.7), (4.10), and (4.12) in the funnel $E_{\hat{\eta},\hat{\xi}}$ is always guaranteed (the same conclusion for the Comb-drive finger model). Since under the same assumption over V_M there is a second pair of constant and positive upper and lower solutions $\tilde{\xi}$ and $\tilde{\eta}$ respectively, but now in reversed order, i.e., $\tilde{\xi} < \tilde{\eta}$, it is necessary to add more conditions. Therefore, it is not difficult to verify that the assumptions (†) and (††) in Theorem 4.11 correspond to those in Theorem 4.2 and Theorem 4.3 for prove the existence of at least a T-periodic solution in the funnel $E_{\tilde{\xi},\tilde{\eta}}$.

4.5.2 *Multiplicity and Stability of Periodic Solutions*

In [22, 23, 38] we found connections between multiplicity and stability of periodic solutions of Liénard equations using degree theory. The following lines are devoted to showing some of these connections. First, we need the following comparison result on Hill's equations.

Lemma 4.4 *Let $q, Q \in C(\mathbb{R}/T\mathbb{Z})$ and $\varrho \geq 0$ such that* [2]

$$q(t) \ll Q(t) \ll \left(\frac{\pi}{T}\right)^2 + \frac{\varrho^2}{4}, \qquad \forall t \in \mathbb{R}.$$

Then the equations

$$(\star) \quad \ddot{v} + \varrho\dot{v} + q(t)v = 0, \quad \textit{and} \quad (\star\star) \quad \ddot{v} + \varrho\dot{v} + Q(t)v = 0,$$

cannot have non-trivial T-periodic solutions simultaneously.

If $\varrho > 0$, the proof of Lemma 4.4 can be found in [38]. For the case $\varrho = 0$, the proof can be achieved using some properties of the rotation $R[a]$ for a Hill's equation

$$\ddot{v} + a(t)v = 0,$$

given in Lemma 3.1 and Lemma 3.2 in [39]. In such a case, if we assume by contradiction that there exist two non-trivial T-periodic solutions of $(\star)$ and $(\star\star)$,

[2] $f \ll g$ means that $f(t) \leq g(t)$ for all t and $f(t) < g(t)$ holds for t in some subset of positive measure.

both equations have Floquet multiplier 1. Then, the corresponding rotation numbers $R[q]$ and $R[Q]$ satisfy

$$R[q] < R[Q] < R[\pi^2/T^2] = \frac{\pi}{T}, \quad \text{and} \quad R[q] = \frac{m_1\pi}{T}, \quad R[Q] = \frac{m_2\pi}{T},$$

for some positive integers, m_1, m_2, which is a contradiction.

With an appropriate application on Lemma 4.4 the authors in [11–14, 16, 17] was able to prove that there are only two positive periodic solutions for the canonical Nathanson model or the graphene-based Nathanson model. The idea is the following: Assume by contradiction that there are three positive periodic solutions $\psi_i(t)$, $i = 1, 2, 3$ of (4.10). In consequence the continuous T-periodic function $z_{j,i}(t) = \psi_j(t) - \psi_i(t)$, $i \neq j$ satisfies the Hill's equation

$$\ddot{x} + \mu\dot{x} + a_{j,i}(t)x = 0,$$

with

$$a_{j,i}(t) = \begin{cases} \frac{g_\alpha(t,\psi_j(t)) - g_\alpha(t,\psi_i(t))}{\psi_j(t)-\psi_i(t)}, & \text{if} \quad \varphi_j(t) \neq \varphi_i(t), \\ \frac{\partial g_\alpha}{\partial x}(t, \psi_i(t)), & \text{if} \quad \psi_j(t) = \varphi_i(t), \end{cases}$$

where $g_\alpha(t, x) = \dfrac{\phi_\alpha(x) - eV^2(t)}{(1-x)^2}$. Notice that

$$(\underline{*}) \qquad 1 - 2eV_m^2 \leq \left(\frac{\pi}{T}\right)^2 + \frac{\mu^2}{4} \quad \Rightarrow \quad a_{j,i}(t) \leq \left(\frac{\pi}{T}\right)^2 + \frac{\mu^2}{4}.$$

Then by Lemma 7.2 in [38] the function $z_{j,i}$ never vanishes, i.e., $z_{j,i}(t) > 0$ or $z_{j,i}(t) < 0$ for all $t \in \mathbb{R}$. This implies that $\psi_i(t)$, $i = 1, 2, 3$ are ordered, without loss of generality let us assume

$$\psi_1(t) < \psi_2(t) < \psi_3(t), \qquad \forall t \in \mathbb{R}.$$

Finally, since $\dfrac{\partial^2 g_\alpha(t, x)}{\partial x^2} < 0$ for all $x \in]0, s_\alpha[$ then $a_{2,1}(t) < a_{3,2}(t)$ or $a_{3,2}(t) < a_{2,1}(t)$. The contradiction appears if we applied Lemma 4.4 to the functions

$$q(t) = a_{2,1}(t), \quad \text{and} \quad Q(t) = a_{3,2}(t).$$

Remark 4.10 Up to this point, the previous results are the key tools to provide the existence of exactly two isolated positive T-periodic solutions $\varphi_1(t)$ and $\varphi_2(t)$ of the graphene-based Nathanson model (4.10), satisfying

$$0 < \xi_1 \leq \varphi_1(t) \leq \eta_1 < x_\alpha, \quad \text{and} \quad x_\alpha < \eta_2 \leq \varphi_2(t) \leq \xi_2.$$

It is remarkable that the same ideas apply perfectly if we consider the Comb-drive finger model (4.2) with constant damping force.

Uniqueness Implies Instability In [22, 23] we found a strategy to study analytically the stability properties of a T-periodic function φ for Duffing equations of type

$$(\underline{\diamond}) \qquad \ddot{x} + \varrho\dot{x} + g(t, x) = 0,$$

where $0 \leq \varrho$ and $g : \mathbb{R} \times \mathbb{R} \to \mathbb{R}$, locally Lipschitz continuous with respect to x and T-periodic in time.

Lemma 4.5 (Lemma 3.2 in [22]) *Assume that every T-periodic solution of $(\underline{\diamond})$ is isolated and let $\eta, \xi \in C^2(\mathbb{R})$ be T-periodic lower and upper solutions of $(\underline{\diamond})$ respectively such that $\eta(t) < \xi(t)$ for all $t \in \mathbb{R}$. Then there exists an unstable T-periodic solution ψ of $(\underline{\diamond})$ satisfying $\eta(t) < \psi(t) < \xi(t)$ for all $t \in \mathbb{R}$.*

Remark 4.11 To be precise, the proof of Lemma 4.5 shows the existence of a T-periodic solution φ of $(\underline{\diamond})$ such that $\text{ind}(\varphi) < 0$ where $\text{ind}(\varphi)$ is the index of φ defined by

$$\text{ind}(\varphi) = \deg[I - \mathcal{P}, B].$$

Here $\deg[\cdot]$ refers to *Brouwer degree* of $I - \mathcal{P}$ (I is the identity in $\mathbb{R}^2$ and $\mathcal{P}$ is the Poincaré-Andronov operator) computed in a neighborhood B of $\xi(0) = (\varphi(0), \dot{\varphi}(0))$ not containing other fixed points of $\mathcal{P}$.

Notice that Lemma 4.5 can be extended to Eq. (4.10) if we consider the function

$$g(t,x) = \begin{cases} \frac{\phi_\alpha(\eta_2) - eV^2(t)}{(1-\eta_2)^2}, & \text{if} \quad x < \eta_2, \\ \frac{\phi_\alpha(x) - eV^2(t)}{(1-x)^2}, & \text{if} \quad \eta_2 \leq x \leq \xi_2, \\ \frac{\phi_\alpha(\xi_2) - eV^2(t)}{(1-\xi_2)^2}, & \text{if} \quad \xi_2 < x. \end{cases}$$

In consequence, since (4.10) has exactly two positive T-periodic solutions the T-periodic solution $\varphi_1(t)$ and $\varphi_2(t)$ with $\varphi_2(t)$ lying in $[\eta_2, \xi_2]$ we deduce that it is unstable. Finally, the proof for the locally asymptotical stability property of $\varphi_1(t)$ is a consequence of the Lyapunov-Zukovskii criterion for Hill's equation. The key point here is to find sufficient conditions such that

$$0 < \frac{\partial g_\alpha(t,x)}{\partial x}, \qquad \text{and} \qquad \frac{\partial g_\alpha(t,x)}{\partial x} \leq \left(\frac{\pi}{T}\right)^2 + \frac{\mu^2}{4},$$

for all $(t, x) \in [0, T] \times [\xi_1, \eta_1]$. The second inequality follows by the assumptions in $(\underline{*})$. New conditions are needed to guarantee the first inequality.

4.5.3 Averaging Theory

In [11] we found a brief introduction to the method of Averaging for computing T-periodic solutions of differential equations of the form

$$\dot{\mathbf{x}} = G_0(t, \mathbf{x}) + \varepsilon G_1(t, \mathbf{x}) + \varepsilon^2 G_2(t, \mathbf{x}, \varepsilon), \tag{4.41}$$

as bifurcation from the case $\varepsilon = 0$ to $\varepsilon \neq 0$ sufficiently small, with $\mathbf{X}$ in some open subset D of $\mathbb{R}^n$, and the functions $G_0, G_1 : \mathbb{R} \times D \to \mathbb{R}^n$ are in the class C^2, T-periodic in t, and for some $\varepsilon_0 > 0$, $G_2 : \mathbb{R} \times D \times]-\varepsilon, \varepsilon[\to \mathbb{R}^n$ is a C^2- function also T-periodic in t. In what follows, the main assumption is that the unperturbed differential equation

$$\dot{\mathbf{x}} = G_0(t, \mathbf{x}), \tag{4.42}$$

has a submanifold of periodic solutions.

Let $\mathbf{x}(t, \mathbf{z})$ be a periodic solution of (4.42) with initial condition $\mathbf{x}(0, \mathbf{z}) = \mathbf{z}$, and consider the linearization of (4.42) along the period solution $\mathbf{x}(t, \mathbf{z})$ given by

$$\dot{\mathbf{y}} = D_{\mathbf{x}} G_0(t, \mathbf{x}(t, \mathbf{z}))\mathbf{y}. \tag{4.43}$$

From now on, Let $M_z(t)$ be some fundamental matrix of the linear differential system (4.43) and $\mathcal{S} : \mathbb{R}^r \times \mathbb{R}^{n-r} \to \mathbb{R}^r$, $\mathbf{x} = (x_1, x_2, \ldots, x_n) \to \mathcal{S}(\mathbf{x}) = (x_1, x_2, \ldots, x_r)$, i.e., $\mathcal{S}$ is the projection of $\mathbb{R}^n$ onto its first r-coordinates. Theorem 4.12 setup conditions for the existence of T-periodic solutions of the perturbed differential system (4.41).

Theorem 4.12 (Theorem 3.1 in [29]) *Let $W \subset \mathbb{R}^r$ be open and bounded, and let $\beta_0 : \overline{V} \to \mathbb{R}^{n-r}$ be a C^2 function. We assume that*

(†) $\mathcal{Z} = \left\{\mathbf{z}_\alpha = (\alpha, \beta_0(\alpha)),\ \alpha \in \overline{W}\right\} \subset D$ *and that for each* $\mathbf{z}_\alpha \in \mathcal{Z}$ *the solution* $x(t, \mathbf{z}_\alpha)$ *of* (4.42) *is T-periodic.*

(†) *For each* $\mathbf{z}_\alpha \in \mathcal{Z}$ *there is a fundamental matrix* $M_{z_\alpha}(t) = M(t, \mathbf{z}_\alpha)$ *of* (4.43) *such that the matrix* $M_{\mathbf{z}_\alpha}^{-1}(0) - M_{\mathbf{z}_\alpha}^{-1}(T)$ *has in the right up corner the* $r \times (n-r)$ *zero matrix, and the right down corner a* $(n-r) \times (n-r)$ *matrix* Δ_α *with* $\det(\Delta_\alpha) \neq 0$.

Let us consider the function $\mathcal{F} : \overline{W} \to \mathbb{R}^r$

$$\mathcal{F}(\alpha) = \mathcal{S}\Big(\int_0^T M_{\mathbf{z}_\alpha}^{-1}(t) G_1(t, \mathbf{x}(t, \mathbf{z}_\alpha))\, dt\Big).$$

If there exists $\alpha^* \in W$ *with* $\mathcal{F}(\alpha^*) = 0$ *and* $\det((d\mathcal{F}/d\alpha)(\alpha^*)) \neq 0$, *then there is a T-periodic solution* $\mathbf{x}(t, \varepsilon)$ *of the system* (4.41) *such that* $\mathbf{x}(0, \varepsilon) \to \mathbf{z}_{\alpha^*}$ *as* $\varepsilon \to 0$.

Next, we present a direct application of Theorem 4.12 for the problem of the bifurcation of T-periodic solutions from the periodic solutions $\mathbf{x}(t, \mathbf{z})$ of (4.42). The key assumption is the existence of an open set V with $\overline{V} \subset D$ such that for each $\mathbf{z} \in \overline{V}$, $\mathbf{x}(t, \mathbf{z})$ is T-periodic, i.e., $\overline{V}$ is formed only by periodic orbits, all of them having the same period T, namely, $\overline{V}$ is *isochronous* for the system (4.42).

Corollary 4.2 (Corollary 1 in [29] Perturbations of an Isochronous Set of Periodic Solutions) *We assume that there exists an open and bounded set V with $\overline{V} \subset D$ such that for each $z \in \overline{V}$, the solution $\mathbf{x}(t, \mathbf{z})$ is T-periodic, then we consider the function $\mathcal{F} : \overline{V} \to \mathbb{R}^n$*

$$\mathcal{F}(\mathbf{z}) = \int_0^T M_{\mathbf{z}}^{-1}(t, \mathbf{z}) G_1(t, \mathbf{x}(t, \mathbf{z}))\, dt.$$

If there exits $\mathbf{z}^* \in W$ with $\mathcal{F}(\mathbf{z}^*) = 0$ and $\det((d\mathcal{F}/d\mathbf{z})(\mathbf{z}^*)) \neq 0$, then there is a T-periodic solution $\mathbf{x}(t, \varepsilon)$ of the system (4.41) such that $\mathbf{x}(0, \varepsilon) \to \mathbf{z}^*$ as $\varepsilon \to 0$.

In [11] the authors applied successfully Corollary 4.2 to the Nathanson model (4.28), or equivalently to the differential system

$$\dot{x} = y, \qquad \dot{y} = -x - \mathrm{e}\Big(\dot{x} - \frac{eV^2(t)}{(1-x)^2}\Big),$$

identifying $\mathbf{x} = (x, y)$ with

$$G_0(t, \mathbf{x}) = (y, -x), \quad G_1(t, \mathbf{x}) = \Big(0, \dot{x} - \frac{eV^2(t)}{(1-x)^2}\Big), \quad \text{and} \quad G_2(t, \mathbf{x}) = (0, 0).$$

Notice that the solutions $\mathbf{x}(t, \mathbf{z}) = (x(t, \mathbf{z}), y(t, \mathbf{z}))$, with $\mathbf{z} = (x_0, y_0)$ of the unperturbed system

$$\dot{x} = y, \qquad \dot{y} = -x,$$

are 2π-periodic functions, moreover are given by

$$x(t, \mathbf{z}) = x_0 \cos t + y_0 \sin t, \qquad y(t, \mathbf{z}) = y_0 \cos t - x_0 \sin t,$$

allowing the application of Corollary 4.2 directly.

4.5.4 Ortega's Principle

We finish this section by showing a novel technique developed by R. Ortega in [31] for the existence of symmetric periodic solutions for a large class of nonlinear oscillators, using two main tools: the shooting method and Sturm comparison and oscillation theory. The class of equation considered in [31] is the following

$$\ddot{x}(t) + x(t)D(t, x(t)) = 0, \tag{4.44}$$

where $D = D(t, x) \in C^{0,1}([0, L]\times\mathbb{R})$ for some $L > 0$ and $C^{0,1}([0, L]\times\mathbb{R})$ denotes the class of functions such that $\partial_x D(t, x)$ exists everywhere with $(t, x) \to D(t, x)$ and $(t, x) \to \partial_x D(t, x)$ continuous on $[0, L] \times \mathbb{R}$. In addition, we have two major assumptions,

$$1.\quad D(t,x) < D(t,0), \quad z \neq 0, \qquad 2.\quad |D(t,x)| \leq \frac{\tilde{C}}{1+|x|}, \quad (t,x) \in [0,L]\times\mathbb{R}, \tag{4.45}$$

where $\tilde{C} > 0$ is a fixed constant. A closer reading of [31] shows us that condition 2. in (4.45) can be replaced by the following one: There exists a positive constant C, such that

$$2'.\quad |xD(t,x)| \leq C, \quad (t,x) \in [0,L]\times\mathbb{R}.$$

A direct computation shows that the condition 2. (or 2′.) implies that all the solutions of (4.44) are well-defined in $[0, L]$ and for a given $v \in \mathbb{R}$, if $x(t, v)$ denote the solution of (4.44) that satisfies the initial conditions $x(0) = 0$, $\dot{x}(0) = v$ then

(i) $x(t, 0) = 0$ for all $t \in \mathbb{R}$.
(ii) The uniqueness of solutions of the Cauchy problem associated to (4.44) implies that each zero of $x(\cdot, v)$ is simple if $v \neq 0$. In consequence, the number of zeros of $x(t, v)$ in $]0, L[$ is finite for each $v \neq 0$ fixed. Therefore, it is possible to define the integer value function $\mathbb{R} \to \mathbb{Z}$, $v \to \nu(v)$

$$\nu(v) = \#\big(\{t \in]0, L[: x(t, v) = 0\}\big).$$

Now consider the variational equation associated with (4.44) at $x = 0$, given by

$$\ddot{z} + D(t, 0)z = 0. \tag{4.46}$$

If $\tilde{z}(t)$ denote the solution of (4.46) that satisfies the initial conditions $z(0) = 0$, $\dot{z}(0) = 1$ define

$$\nu_0 = \#\big(\{t \in]0, L[: \tilde{z}(t, v) = 0\}\big).$$

In [31] are proved the following properties:

(I) $\nu(v) \leq \nu_0$ for each $v \neq 0$.

(II) Given $\omega \neq 0$ there exists $\delta > 0$ such that $\nu(\omega) \leq \nu(v) \leq \nu(\omega) + 1$ if $|v - \omega| \leq \delta$. Moreover, if $x(L, \omega) \neq 0$ then $\nu(v) = \nu(\omega)$ for $|v - \omega| \leq \delta$.
(III) There exists $v_* > 0$ such that $\nu(v) = \nu_0$ if $|v| \leq \nu_*$.
(IV) There exists $v^* > 0$ such that $\nu(v) = 0$ if $|v| \geq \nu^*$.

The combination of properties $I)$ to $IV)$ provide the elements for the proof of Theorem 4.13 (Proposition 4 in [31]).

Theorem 4.13 (Ortega's Principle) *Let us consider the Dirichlet problem*

$$\ddot{x}(t) + x(t)D(t, x(t)) = 0, \quad x(0) = 0, \quad \dot{x}(L) = 0, \tag{4.47}$$

where $D = D(t, x)$ *satisfies the conditions* (4.45). *Then,* (4.47) *has a solution with* $N \geq 0$ *zeros in* $]0, L[$ *if and only if* $\nu_0 > N$.

The following additional and very interesting result is given in [31] for the existence of positive solution of (4.47), i.e., $x(t) > 0$ for all $t \in]0, L[$.

Theorem 4.14 (Proposition 5 in [31]) *Assume that* D *is a function of class* $C^{0.1}([0, L] \times \mathbb{R})$ *satisfying*

$$\partial_x D(t, x) < 0, \quad \textit{for each } t \in [0, L] \textit{ and } x > 0.$$

Then the Dirichlet problem (4.47) *has at most one positive solution.*

Remark 4.12 We invited the reader to see in [31] how Theorem 4.13 and Theorem 4.14 emerge and was successfully applied to the problem of the existence of odd and $2m\pi$-periodic solutions with a prescribed number of zeros in $[0, m\pi]$ (for each integer $m \geq 1$), of the restricted three-body problem known nowadays as the Sitnikov problem (see [31] and the references therein) and given by

$$\ddot{x}(t) + \frac{x}{(r^2(t, \varsigma) + x^2(t))^{3/2}} = 0,$$

where $r(t, \varsigma)$ is the distance of the two bodies with equal mass (called primaries) lying in the same plane Π, moving in elliptic orbits (with eccentricity $\varsigma \in [0, 1[$) around the center of mass (barycenter) of the 2-body problem and the massless body moves in the x-axis which is orthogonal to the plane Π and passing through the barycenter of the primaries.

4.6 Some Open Problems

In view of the results and discussions in Sect. 4.3 related to the search for periodic solutions of the Nathanson model and the Comb-drive finger model, we end this review by pointing out some open mathematical problems.

With respect to the Nathanson model (4.7), in [40] the author states the following:

Open Problem 1 *To prove the existence of a saddle-node bifurcation of periodic solutions for* (4.7) *with* $V_m \leq 0$*, i.e., when* $\delta \geq -V_0/v_m$.

As the reader can deduce, Open Problem 1 call for a result like Theorem 4.1 on the existence, multiplicity, and stability of periodic solutions of (4.7) but for a function $V(t)$ that may change sign. To our knowledge, [41] is the first paper to consider mathematically and numerically the Open Problem 1. Here, the authors argue to prove mathematically the existence of at least one periodic solution of (4.7) (see Theorem 2.1 in there). But, a careful reading of their arguments shows that the proof has a critical failure. In addition, using numerical bifurcation theory, the authors in [41] provide an answer to Open Problem 1.

Open Problem 2 Extend or obtain new results of periodic solutions for all the nonlinear oscillators showed in Sect. 4.3 for a function $V(t)$ that may change sign.

The Open Problem 2 considers for example the analytical and numerical analysis of periodic motions of the nonlinear oscillator

$$\ddot{x}(t) + \frac{\gamma}{(1-x(t))^3}\dot{x}(t) + h(x(t)) = \frac{eV^2(t)}{(1-x(t))^2},$$

with $e, \gamma > 0$, $V(t) \in C(\mathbb{R}/T\mathbb{Z})$, and $h(x)$ a real value function that satisfies some appropriate conditions. For example in [16] we found

1. $h \in C^1(\mathbb{R}, \mathbb{R})$ and is an odd function.
2. $\exists x_* > 0 : h(x_*) = 0$, $h \in C^2(]0, x_*[, \mathbb{R})$ concave and positive on $]0, x_*[$, and satisfies $h'(x_*) < 0 < h'(0)$.

Another important class of electromechanical systems where a Liénard type equation with two singularities governs the motion of the moveable parts is *Torsional Actuator* (see [4, 42]). In general, a parallel plate torsional actuator is composed of two electrodes. One electrode (moveable) is connected to two torsional springs and the second one (fixed) is placed underneath a fixed distance on the substrate (see Fig. 4.10). In appropriate units, the electrostatic torque $\theta = \theta(t)$ satisfies the equation (see [42])

$$\ddot{\theta}(t) + \gamma\dot{\theta}(t) + \theta(t) = \frac{bV^2(t)}{\theta^2(t)}\left(\frac{(\nu_2-\nu_1)\theta(t)}{(1-\nu_2\theta(t))(1-\nu_1\theta(t))} + \ln\left(\frac{1-\nu_2\theta(t)}{1-\nu_1\theta(t)}\right)\right) + f(\theta(t)), \tag{4.48}$$

where $\gamma, b \in \mathbb{R}^+$ and $0 \leq \nu_1 < \nu_2$, are design and construction parameters and $V = V(t)$ is the applied voltage load and $f(\theta)$ the intermolecular forces between the electrodes[3]. The two intermolecular forces commonly considered are the Van

[3] When intermolecular forces appear, the dimension of the device is in nanometers is such case we are dealing with nano-electromechanical systems (NEMS)

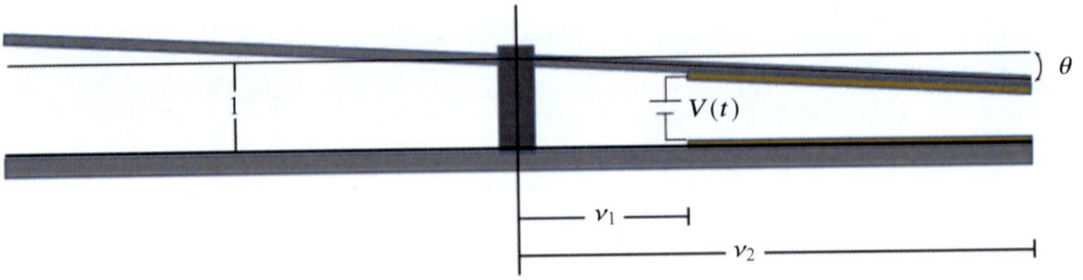

Fig. 4.10 Side view of the rotational parallel plate actuator. The torsional angle $\theta = \theta(t)$ measures the angular deviation of the moveable electrode

der Waals force $f_{v_w}(\theta)$ given by

$$f_{v_w}(\theta) = \frac{\sigma_{v_w}}{(1-\theta)^2}, \qquad \sigma_{v_w} \geq 0,$$

or the Casimir force $f_{c_m}(\theta)$,

$$f_{c_m}(\theta) = \frac{\sigma_{c_m}(a-\theta)}{(1-\theta)^3}, \qquad a > 1,\ \sigma_{c_m} \geq 0.$$

In the particular case when the fixed electrode extends the whole length of the actuator, i.e., $\nu_1 = 0$, Eq. (4.48) reduces to

$$\ddot{\theta}(t) + \gamma\dot{\theta}(t) + \theta(t) = \frac{b\eta\mathcal{V}^2(t)}{\theta^2(t)}\left(\frac{\nu_2\theta(t)}{1-\nu_2\theta(t)} + \ln(1-\nu_2\theta(t))\right) + f(\theta(t)). \tag{4.49}$$

Open Problem 3 *To study the dynamics and the existence of periodic solutions for the rotational actuator* (4.48) *(or* (4.49)*) under a periodic voltage source* $V(t)$ *and intermolecular force* $f(\theta)$ *like a Van der Waals force or Casimir force.*

Finally, if intermolecular forces between the electrodes in the canonical two-parallel plate actuator are considered, it can be proved (see [42]) that the equation of motion of the moveable electrode under Van der Waals force is given by

$$\ddot{x}(t) + c(x(t))\dot{x}(t) + x(t) = \frac{eV^2(t)}{(1-x(t))^2} + \frac{\delta_{v_w}}{(1-x(t))^3}, \tag{4.50}$$

and for Casimir force

$$\ddot{x}(t) + c(x(t))\dot{x}(t) + x(t) = \frac{eV^2(t)}{(1-x(t))^2} + \frac{\delta_{c_m}}{(1-x(t))^4}, \tag{4.51}$$

with $e, \delta_{v_w}, \delta_{c_m} > 0$.

Open Problem 4 *To study analytically and numerically the dynamics and the existence of periodic solutions of parallel plate actuator* (4.50) *or* (4.51) *under a viscous damping* ($c(x) = c$, $c > 0$) *or a squeeze-film damping* ($c(x) = \mu/(1-x)^3$, $\mu > 0$) *and a voltage source* $V(t) = V_0 + \delta v(t)$, *where* $V_0, \delta \in \mathbb{R}^+$ *and* $v \in C(\mathbb{R}/T\mathbb{Z})$ *for some* $T > 0$, *and with zero average, i.e.,* $\int_0^T v(t)\,dt = 0$.

These are just few of the challenging problems in the analysis of the dynamics of some fundamental MEMS and NEMS devices that continuously draw the attention of several scientists.

References

1. H. Nathanson, W. Newell, R. Wickstrom, J. Davis: The resonant gate transistor. IEEE Trans. Electron Dev. **14**(3), 117–133 (1967)
2. P. Laurencot, C. Walker, Some singular equations modeling mems. Bull. Am. Math. Soc. **54**, 437–479 (2016)
3. J.A. Pelesko, D.H. Bernstein, *Modeling MEMS and NEMS* (CRC Press, 2002)
4. M. Younis, *MEMS Linear and Nonlinear Statics and Dynamics* (Microsystems, Springer US, 2011)
5. J.G. Korvink, P. Oliver, *MEMS: A Practical Guide to Design, Analysis, and Applications* (Springer, Berlin, Heidelberg, 2006)
6. D. Rafiq, M.A. Bazaz, A collection of large-scale benchmark models for nonlinear model order reduction. Arch. Comput. Methods Eng. **30**(1), 69–83 (2022)
7. K. Pyragas, Continuous control of chaos by self-controlling feedback. Phys. Lett. A **170**(6), 421–428 (1992)
8. F.M. Alsaleem, M.I. Younis, H.M. Ouakad, On the nonlinear resonances and dynamic pull-in of elec- trostatically actuated resonators. J. Micromech. Microeng. **19**(4), 045013 (2009)
9. F.M. Alsaleem, M.I. Younis, Stabilization of electrostatic mems resonators using a delayed feedback controller. Smart Mater. Struct. **19**(3), 035016 (2010)
10. S. Shao, K.M. Masri, M.I. Younis, The effect of time-delayed feedback controller on an electrically actuated resonator. Nonlinear Dynam. **74**(1), 257–270 (2013)
11. J. Llibre, D. Nuñez, A. Rivera, Periodic solutions of the nathanson's and the comb-drive models. Int. J. Non-Linear Mech. **104**, 109–115 (2018)
12. D. Núñez, O. Perdomo, A. Rivera, On the stability of periodic solutions with defined sign in mems via lower and upper solutions. Nonlinear Anal. Real World Appl. **46**, 195–218 (2019)
13. J. Beron, A. Rivera, Periodic oscillations in mems under squeeze film damping force. J. Appl. Math. **2022**, 1–15 (2022)
14. A. Gutíerrez, P. Torres, Non-autonomous saddle-node bifurcation in a canonical electrostatic mems. Int. J. Bifurcat. Chaos **23**(05), 1–9 (2013)
15. D. Wei, S. Kadyrov, Z. Kazbek, Periodic solutions of a graphene-based model in micro-electro-mechanical pull-in device. Appl. Comput. Mech. **11**(1) (2017). https://doi.org/10.24132/acm.2017.322
16. D. Núñez, L. Murcia, J. Galán, Stable periodic oscillations in simple parallel-plate mems based on a family of graphene-like materials. Int. J. Non-linear Mech. **149** (2023). https://doi.org/10.1016/j.ijnonlinmec.2022.104324
17. A. Rivera, J. Thibodeaux, J.S. Sánchez, Existence and stability of periodic solutions of a shifted comb-drive finger actuator. J. Appl. Nonlinear Dynam. **11**(1), 247–269 (2022)
18. W.-M. Zhang, H. Yan, Z.-K. Peng, G. Meng, Electrostatic pull-in instability in mems/nems: A review. Sens. Actuat. A Phys. **214**, 187–218 (2014)
19. E. Picard, Sur les méthodes d'approximations successives dans la théorie des équations différentielles. Am. J. Math. **20**(1), 87–100 (1898)

20. G.S. Dragoni, Il problema dei valori ai limiti studiato in grande per le equazioni differenziali del secondo ordine. Mathematische Annalen **105**(1), 133–143 (1931)
21. R. Ortega, Stability and index of periodic solutions of an equation of duffing type. Boll. Unione Mat. Ital. **7**, 533–546 (1989)
22. R. Ortega, Topological degree and stability of periodic solutions for certain differential equations. J. Lond. Math. Soc. **42**, 505–516 (1990)
23. E. Dancer, R. Ortega, The index of lyapunov stable fixed points in two dimensions. J. Dynam. Differ. Equ. **6**, 631–637 (1994)
24. R. Ortega, Some applications of the topological degree to stability theory, in *Topological Methods in Differential Equations and Inclusions*, ed. by A. Granas, M. Frigon. NATO ASI Series, vol. 472, 377–409 (1995)
25. S. Kadyrov, et al., Periodic solutions and the avoidance of pull-in instability in nonautonomous microelectromechanical systems. Math. Methods Appl. Sci. **44**, 14556–14568 (2021)
26. J.P. Berón, Existence, Uniqueness, and Stability of Periodic Solutions of Lienard type Equations. Disset, Bachelor's Thesis. Pontificia Universidad Javeriana Cali, Facultad de Ingeniería y Ciencias (2020)
27. P. Amster, A. Rivera, J.A. Arredondo, Periodic oscillations in electrostatic actuators under time delayed feedback controller. Commun. Nonlinear Sci. Numer. Simul. **131**, 107840 (2024). https://doi.org/10.1016/j.cnsns.2024.107840
28. W. Magnus, S. Winkler., *Hill's Equation* (Dover Publications, 1979)
29. A. Buica, J. Francoise, J. Llibre, Periodic solutions of nonlinear differential systems with a small parameter. Commun. Pure Appl. Anal. **6**(1), 103–111 (2007)
30. D. Núñez, O. Larreal, L. Murcia, Odd periodic oscillations in comb-drive finger actuators. Nonlinear Anal. Real World Appl. **61**, 103347 (2021)
31. R. Ortega, Symmetric periodic solutions in the sitnikov problem. Archiv der Mathematik **107**, 405–416 (2021)
32. P. Amster, *Topological Methods in the Study of Boundary Value Problems* (Springer, 2014)
33. A. Gutierrez, D. Núñez, A. Rivera, Effects of voltage change on the dynamics in a comb-drive finger of an electrostatic actuator. Int. J. Non-Linear Mech. **95**, 224–232 (2017)
34. R. Gaines, A priori bounds and upper and lower solutions for nonlinear second-order boundary-value problems. J. Differ. Equ. **12**(2), 291–312 (1972)
35. R. Hakl, P.J. Torres, M. Zamora, Periodic solutions of singular second order differential equations: Upper and lower functions. Nonlinear Anal. Theory Methods Appl. **74**(18), 7078–7093 (2011)
36. M. Nagumo, über die differentialgleichung $y'' = f(x, y, y')$. Proc. Phys. Math. Soc. Jpn. 3 **19**, 861–866 (1937)
37. D.C. Coster, P. Habets, *Two-Point Boundary Value Problems Lower and Upper Solutions* (Elsevier, 2006)
38. J. Mawhin, Topological degree and boundary values problems for nonlinear differential equations. *Lecture notes in Mathematics* (vol. 1537(1)) (Springer, Berlin, Heidelberg, 1993), pp. 74–142
39. J. Lei, X. Li, P. Yan, M. Zhang, Twist character of the least amplitude periodic solution of the forced pendulum. SIAM J. Math. Anal. **35**(4), 844–867 (2003)
40. P.J. Torres, *Mathematical Models with Singularities* (Atlantis Press, Paris, 2015)
41. Q. Yuan, Z. Cheng, X. Li, Dynamics of periodic solution to a electrostatic micro-electromechanical system. Commun. Nonlinear Sci. Numer. Simul. **116** (2023). https://doi.org/10.1016/j.cnsns.2022.106828
42. W.-H. Lin, Y.-P. Zhao, Influence of damping on the dynamical behavior of the electrostatic parallel-plate and torsional actuators with intermolecular forces. Sensors **7**(12), 3012–3026 (2007)

Chapter 5
Analysis of a Mathematical Model of Competition in a Chain of Periodic Chemostats in Series

Guillermo Contreras, Gonzalo Robledo, and Daniel Sepúlveda

MSC Classification 34K20, 92D25, 393D05

5.1 Introduction

5.1.1 Preliminaries

The chemostat [14, 25, 26, 34] is a continuous bioreactor where one or more microbial species are cultivated in a well stirred liquid medium, which contains the nutrients necessary to ensure the microbial species growth; with the exception of one specific nutrient, which is called the *limiting nutrient* and is pumped into the vessel from an external source. The first part of this paper considers a chemostat containing two species in competition for the limiting nutrient, whose dynamics are described by the following periodic ODE system:

$$\begin{cases} \dot{s} = D[s^0 + b\,e(t) - s] - f_1(s)x_1 - f_2(s)x_2, \\ \dot{x}_1 = x_1[f_1(s) - D], \\ \dot{x}_2 = x_2[f_2(s) - D], \end{cases} \tag{5.1}$$

All the authors contributed equally to this work.

G. Contreras
Colegio Pedro de Valdivia, Quilín 5473, Peñalolén–Santiago, Chile
e-mail: gcontreras@cpdv.cl

G. Robledo (✉)
Departamento de Matemáticas, Universidad de Chile, Ñuñoa–Santiago, Chile
e-mail: grobledo@uchile.cl

D. Sepúlveda
Departamento de Matemática, Universidad Tecnológica Metropolitana, Ñuñoa–Santiago, Chile
e-mail: daniel.sepulveda@utem.cl

P. Amster, P. Benevieri (eds.), *Topological Methods for Delay and Ordinary Differential Equations*, Advances in Mechanics and Mathematics 51,
https://doi.org/10.1007/978-3-031-61337-1_5

where x_1 and x_2 denote the densities of two microbial species at time t, while s denotes the concentration of the limiting nutrient at time t. Throughout this work, it will be assumed that any initial condition is componentwise positive.

The bioreactor receives an input of liquid volume at rate $D > 0$, which contains a concentration of limiting nutrient described by the T-periodic and continuous function $t \mapsto s^0 + be(t)$, where s^0 and b are positive constants and $t \mapsto e(t)$ is a continuous and T-periodic function such that $t \mapsto s^0 + be(t)$ is also positive. Simultaneously, a liquid volume containing a mixture of biomass and limiting nutrient is pumped out of the bioreactor, at the same rate $D > 0$, which ensures a constant volume.

A hypothesis critical of several chemostat models states that the consumption of the limiting nutrient has a direct relation with the per-capita growth of the microbial species, which is mediated by the *nutrient uptake rates* $f_1(s)$ and $f_2(s)$, the so called uptake functions.

The system (5.1) is a particular case of the model introduced by J. Hale and A. Somolinos in [11] for $n \geq 2$ microbial species, for which sharp results were obtained under the following assumption:

(H1) The functions $f_i\colon [0, +\infty) \to [0, +\infty)$ are described by:

$$f_i(s) = m_i \frac{s}{a_i + s}, \tag{5.2}$$

where the positive constants m_i and a_i are named respectively the *maximal growth rate* and the *half-saturation constant*, both associated with the i-th species. This function is usually called the *Michaelis–Menten* function or *Monod's* function and is ubiquitous in the bio-processes literature; we would like to refer the reader to [14, 23, 24, 34] for details.

Now, we will point out two prominent facts treated in [11]: first of all, notice that the change of variables $z = s + x_1 + x_2$ leads to the scalar nonautonomous differential equation:

$$\dot{z} = -Dz + D[s^0 + be(t)]. \tag{5.3}$$

Second, it helps to define the auxiliary functions $g_j\colon [0, +\infty) \to \mathbb{R}$ given by:

$$g_j(s) := f_j(s) - D \quad \text{for } j = 1, 2. \tag{5.4}$$

In fact, notice that the j-th microbial species will have a positive growth if and only if the limiting nutrient inside the bioreactor is such that $g_j(s) > 0$. This premise prompts us to introduce the notion of *break-even* concentrations of the j-th species as:

$$\lambda_j = g_j^{-1}(0) = \frac{a_j D}{m_j - D}, \tag{5.5}$$

which describes the minor concentration of limiting nutrient required by the j-th species in order to have a positive growth.

Proposition 5.1 *The differential equation (5.3) has a unique T-periodic solution*

$$\Sigma_0^*(t) = \int_{-\infty}^{t} De^{-D(t-s)}[s^0+be(s)]\,ds = s^0+b\int_{-\infty}^{t} De^{-D(t-s)}e(s)\,ds, \qquad (5.6)$$

which is also positive, continuous, and globally attractive; that is, any solution $t \mapsto \Sigma_0(t)$ of (5.3) verifies the asymptotic behavior:

$$\lim_{t\to+\infty} \mid \Sigma_0(t) - \Sigma_0^*(t) \mid = 0. \qquad (5.7)$$

Proof By Theorem 2.1 from [40], we can deduce that $t \mapsto \Sigma_0^*(t)$ is the unique T-periodic solution of (5.3). Moreover, if $t \mapsto \Sigma(t)$ is another solution of (5.3), then we define $u(t) = \Sigma_0^*(t) - \Sigma(t)$ and observe that $\dot{u} = -Du$. Then we have

$$\mid \Sigma_0^*(t) - \Sigma(t) \mid = \mid u(0) \mid e^{-Dt},$$

and the limit (5.7) follows. □

The above Proposition allows us to construct a recursive sequence of T-periodic functions defined as follows:

$$\Sigma_n^*(t) = \int_{-\infty}^{t} De^{-D(t-s)}\Sigma_{n-1}^*(s)\,ds, \qquad (5.8)$$

where $t \mapsto \Sigma_n^*(t)$ is the unique T-periodic solution of

$$\dot{\Sigma}_n = -D\Sigma_n + D\Sigma_{n-1}^*(t),$$

while any solution $t \mapsto \Sigma_n(t)$ of the above differential equation also verifies

$$\lim_{t\to+\infty} \mid \Sigma_n(t) - \Sigma_n^*(t) \mid = 0. \qquad (5.9)$$

Despite the above sequence has interest on itself, we point out that we will essentially work with the terms $n = 0, 1, 2$. The term $n = 0$ prompts us to define the set

$$\Omega_0 := \left\{(s, x_1, x_2) \in \mathbb{R}_+^3 : 0 < s + x_1 + x_2 < s^0 + be_{\max}\right\}, \qquad (5.10)$$

where $\mathbb{R}_+^3$ is the interior of the positive cone of $\mathbb{R}^3$ and $e_{\max} := \max_{t\in[0,T]} e(t)$.

Lemma 5.1 *Let $t \mapsto u(t) := (s(t), x_1(t), x_2(t))$ be a solution of (5.1) passing through u_0 at $t = t_0$. If $u_0 \in \Omega_0$, then $t \mapsto u(t) \in \Omega_0$ for any $t \geq t_0$.*

Proof Let $t \mapsto (s(t), x_1(t), x_2(t))$ be a solution of (5.1) passing trough $u_0 \in \Omega_0$ at $t = t_0$. As it is straightforward to verify that all the components are positive for any fixed $t > t_0$, we only will prove that $s(t) + x_1(t) + x_2(t) < s^0 + be_{\max}$ for any fixed $t > t_0$.

As we know, the change of variables $z = s + x_1 + x_2$, implies that $t \mapsto z(t) : s(t) + x_1(t) + x_2(t)$ is a solution of the scalar differential equation (5.3) with initial condition $z(t_0) < s^0 + be_{\max}$. Then, we have that

$$\begin{aligned} z(t) &= e^{-D(t-t_0)} z(t_0) + D \int_{t_0}^{t} e^{-D(t-r)} [s^0 + be(r)]\, dr \\ &< [s^0 + be_{\max}] \left\{ e^{-D(t-t_0)} + De^{-Dt} \int_{t_0}^{t} e^{Dr}\, dr \right\} \\ &= s^0 + be_{\max}, \end{aligned}$$

and the Lemma follows. □

5.1.2 *Competitive Exclusion*

The system (5.1) can be considered as a scenario where two microbial species are in competition for the limiting nutrient. In this context, the species requiring the minor amount of nutrient ensuring to have a positive growth seems to have a competitive advantage over the other species. Without loss of generality, we will assume that the break-even concentrations verify the inequalities

$$\lambda_1 < \lambda_2 < s^0, \tag{5.11}$$

that is, the first microbial species needs a minor amount of limiting nutrient than the second one in order to have a positive per-capita growth. This species can be considered as the advantaged competitor.

The asymptotic behavior of the solutions of (5.1) satisfying **(H1)** and (5.11) has been extensively studied; epitomical amongst them is the case $b = 0$, for which we recall the following result from [34, p.15] :

Proposition 5.2 (Competitive Exclusion Principle) *If (5.11) and* **(H1)** *are satisfied, as well as $b = 0$, then any solution of (5.1) with positive initial conditions verifies:*

$$\lim_{t \to +\infty} (s(t), x_1(t), x_2(t)) = (\lambda_1, s^0 - \lambda_1, 0).$$

The above result is an example of the *competitive exclusion principle* applied to a homogeneous chemostat. The competitive exclusion principle is a pivotal topic in theoretical ecology; it states that given two competing species: A and B, in the same niche and geographical area, if the species A grows faster than the species B, then this last species will be displaced and driven to extinction. We point out to [13], and the references therein, for details.

The competitive exclusion principle has been experimentally verified in microbial ecosystems described by the Eq. (5.1), and we refer to [7, 12, 17, 27] for specific examples. Withal, the competitive exclusion principle has also been theoretically verified in more general differential equations describing autonomous chemostats; some of the equations are models considering: (i) not explicit uptake functions [19, 28], (ii) delays between the consumption of limiting nutrient and its effect on microbial growth [10, 20, 21], (iii) mortality rate of the microbial biomass [9, 19, 21], (iv) stochastic perturbations [39, 41], and (v) variable yields [31].

Nevertheless, there is no abundance of mathematical results in the case of the competitive exclusion principle in nonautonomous chemostat equations. The pioneers in this research are J. Hale and A. Somolinos [11], and we will use their results restricted to $n = 2$, where they understood the inequality (5.11) is being satisfied, and focused in finding sufficient conditions described in terms of $f_i(\cdot)$, b, s^0 and $e(t)$ such that the competitive exclusion is still verified in the periodic model (5.1). The following result is not explicitly stated in [11] but can be deduced after a careful reading:

Proposition 5.3 *Let us consider a chemostat (5.1). If*

(a) The break-even concentrations satisfy $\lambda_1 < \lambda_2 < s^0$.
(b) $m_1 < m_2$ *and there exists a constant* B_2 *such that* $s^0 + b < B_2$ *and*

$$B_2 \geq \frac{a_1 a_2(\lambda_2 - \lambda_1) + \lambda_2\lambda_1(a_2 - a_1)}{(a_2\lambda_1 - a_1\lambda_2)} > \lambda_2.$$

(c) The function f_1 *verifies the average condition*

$$\frac{1}{T}\int_0^T f_1(\Sigma_0^*(t))\,dt > D, \qquad (\mathcal{A}_1)$$

then there exists a T*-periodic solution* $t \mapsto (s_1^*(t), x_1^*(t), 0)$, *such that any positive solution* $t \mapsto (s(t), x_1(t), x_2(t))$*of (5.1) verifies*

$$\lim_{t\to+\infty} \mid s_1(t) - s_1^*(t) \mid + \mid x_1(t) - x_1^*(t) \mid + \mid x_2(t) \mid = 0,$$

and the competitive exclusion is still verified.

Proof By the conditions (a) and (b), the Theorem 2.5 from [11] implies that $t \mapsto x_2(t)$ converges exponentially to zero when $t \to +\infty$. Now, the result is a direct consequence of the condition (c) combined with the Corollary 5.6 from [11]. □

Last but not least, it is important to emphasize that the competitive exclusion is an unusual behavior in the nonautonomous context. In fact, there exists a plethora

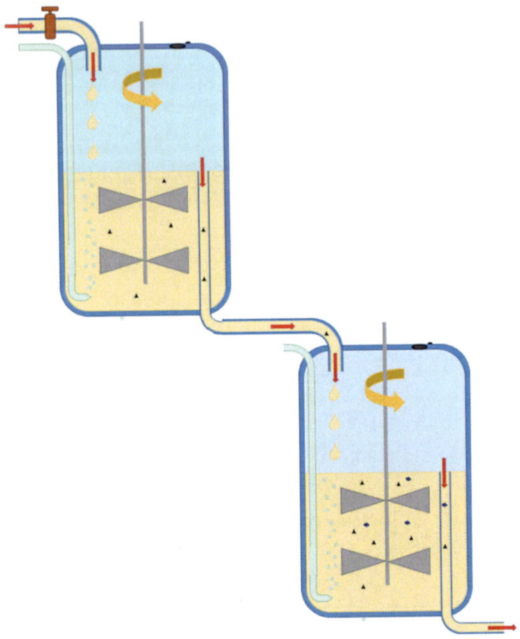

Fig. 5.1 Scheme of the chain of chemostats: The less advantaged competitor (black triangles) is cultivated in the first vessel whose output strengthens its competitive capacities with the more advantaged species (blue points) in the second vessel

of results describing sufficient conditions for the existence of positive T-periodic solutions of (5.1); we refer to [38] and [42, Ch.5] for details.

5.2 Avoiding Competitive Exclusion: Chain of Chemostats

Throughout this article, we will assume that the hypotheses of Proposition 5.3 are satisfied and, in consequence, the competitive exclusion is verified for (5.1). In order to avoid the competitive exclusion, we will construct a chain of two chemostats to promote the coexistence of both microbial species, a diagram of this chain is presented in Fig. 5.1.[1] The chain has the following properties:

- The less advantaged competitor x_2 is cultivated separately in a first chemostat, whose output will be the input of (5.1).
- The density of the less advantaged competitor will be renamed by y_1 and y_2, where y_1 denotes its concentration in the first chemostat while y_2 denotes its

[1] The image of Fig. 5.1 was generated from a file created by CGraham2332 called Chemostat Vessel Diagram.png available at https://commons.wikimedia.org/ under the Creative Commons Attribution-Share Alike 4.0 International license (https://creativecommons.org/licenses/by-sa/4.0/deed.en).

density in the second one. The more advantaged competitor will be denoted by z_2 and will be present only in the second chemostat.

- The concentration of the limiting nutrient will be denoted by s_1 and s_2, where s_1 is the concentration in the first chemostat while s_2 will be the concentration in the second one.

The dynamics in the first chemostat is described for any $t \geq 0$ by the system

$$\begin{cases} \dot{s}_1 = D[s^0 + b\,e(t) - s_1] - f_1(s_1)y_1, \\ \dot{y}_1 = f_1(s_1)y_1 - Dy_1. \end{cases} \tag{5.12}$$

Now, if the output of (5.12) becomes the input of (5.1), we obtain the coupled system:

$$\begin{cases} \dot{s}_1 = D[s^0 + be\,(t) - s_1] - f_1(s_1)y_1, \\ \dot{y}_1 = f_1(s_1)y_1 - Dy_1, \\ \dot{s}_2 = D[s_1 - s_2] - f_1(s_2)y_2 - f_2(s_2)z_2, \\ \dot{y}_2 = f_1(s_2)y_2 + D[y_1 - y_2], \\ \dot{z}_2 = f_2(s_2)z_2 - Dz_2. \end{cases} \tag{5.13}$$

The system (5.13) was introduced by G. Contreras for the autonomous case with $b = 0$ in [6], where the main result is a set of sufficient conditions ensuring the existence and global asymptotic stability of a positive equilibrium. This result has been extended in [22] by considering small delays between consumption of the limiting nutrient and its metabolization by the microbial biomass. To the best of our knowledge, there are no results for this chain in a nonautonomous framework.

5.3 Study of the Chain

5.3.1 Preliminaries

From now on, we will consider initial conditions of the system (5.12) on the set

$$\Omega_1 := \left\{(s_1, y_1) \in \mathbb{R}^2_+ : 0 < s_1 + y_1 < s^0 + be_{\max}\right\}$$

and initial conditions for the system (5.13) on the set:

$$\Omega_2 := \left\{(s_1, y_1, s_2, y_2, z_2) \in \mathbb{R}^5_+ : (s_1, y_2) \in \Omega_1 \quad \text{and} \quad (s_2, y_2, z_2) \in \Omega_0\right\},$$

where Ω_0 is defined by (5.10). Finally, the following forward invariance result can be proved similarly as done in Lemma 5.1:

Lemma 5.2 *Let*

$$t \mapsto u(t) := (s_1(t), y_1(t)) \quad \text{and} \quad t \mapsto v(t) := (s_1(t), y_1(t), s_2(t), y_2(t), z_2(t))$$

be respectively solutions of (5.12) and (5.13) passing through u_0 and v_0 at $t = t_0$. If $u_0 \in \Omega_1$ and $v_0 \in \Omega_2$, then $t \mapsto u(t) \in \Omega_1$ and $t \mapsto v(t) \in \Omega_2$ for any $t \geq t_0$.

Notice that the solutions of the subsystem (5.12) are independent of (s_2, y_2, z_2) and its asymptotic behavior has been studied previously in the literature.

Proposition 5.4 ([11, Th.5.5], [42, Th.5.2.2]) *If (H1) and the average condition $(\mathcal{A}_1)$ are satisfied, then there exists a positive T-periodic solution $t \mapsto (s_1^*(t), y_1^*(t))$ of (5.12) such that any other solution $t \mapsto (s_1(t), y_1(t))$ with positive initial conditions verifies*

$$\lim_{t \to +\infty} \mid s_1(t) - s_1^*(t) \mid + \mid y_1(t) - y_1^*(t) \mid = 0.$$

Remark 5.1 The uniqueness of the T-periodic solution $t \mapsto \Sigma_0^*(t)$ of (5.3) implies that

$$\Sigma_0^*(t) = s_1^*(t) + y_1^*(t) \quad \textit{for any } t \in \mathbb{R}. \tag{5.14}$$

5.3.2 A Modified System

Note that the changes of variables $\Sigma_0 = s_1 + y_1$ and $\Sigma_1 = s_2 + y_2 + z_2$ transform the system (5.13) into:

$$\begin{cases} \dot{\Sigma}_0 = -D\Sigma_0 + D[s^0 + be(t)], \\ \dot{\Sigma}_1 = -D\Sigma_1 + D\Sigma_0, \\ \dot{y}_1 = f_1(\Sigma_0 - y_1)y_1 - Dy_1, \\ \dot{y}_2 = f_1(\Sigma_1 - y_2 - z_2)y_2 + D[y_1 - y_2], \\ \dot{z}_2 = f_2(\Sigma_1 - y_2 - z_2)z_2 - Dz_2. \end{cases} \tag{5.15}$$

In addition, notice that the corresponding set of initial conditions of the system (5.15) is given by

$$W := \left\{ (\Sigma_0, \Sigma_1, y_1, y_2, z_2) \in \mathbb{R}_+^5 : \begin{array}{c} 0 < y_1 < \Sigma_0 < s^0 + be_{\max} \\ \\ 0 < y_2 + z_2 < \Sigma_1 < s^0 + be_{\max} \end{array} \right\}. \tag{5.16}$$

A direct consequence of Lemma 5.2 combined with the definition of the variables Σ_0 and Σ_1 is that any solution $t \mapsto w(t) = (\Sigma_0(t), \Sigma_1(t), y_1(t), y_2(t), z_2(t))$ of

(5.15) passing through $w_0 \in W$ at $t = t_0$ verifies $w(t) \in W$ for any $t \geq t_0$. On the other hand, it will be useful to emphasize that the system (5.15) has the triangular or hierarchical structure

$$\begin{cases} \dot{u} = G(t, u) \\ \dot{v} = F(u, v) \end{cases} \quad \text{or its abbreviated form} \quad \dot{\xi} = Q(t, \xi), \tag{$\mathcal{T}$}$$

where the first subsystem $\dot{u} = G(t, u)$ is given by

$$\begin{cases} \dot{\Sigma}_0 = -D\Sigma_0 + D[s^0 + be(t)], \\ \dot{\Sigma}_1 = -D\Sigma_1 + D\Sigma_0, \\ \dot{y}_1 = f_1(\Sigma_0 - y_1)y_1 - Dy_1, \end{cases} \tag{$\mathcal{G}$}$$

whose solutions are independent of the second subsystem $\dot{v} = F(u, v)$, given by

$$\begin{cases} \dot{y}_2 = f_1(\Sigma_1 - y_2 - z_2)y_2 + D[y_1 - y_2], \\ \dot{z}_2 = f_2(\Sigma_1 - y_2 - z_2)z_2 - Dz_2. \end{cases} \tag{$\mathcal{F}$}$$

Lemma 5.3 *The subsystem ($\mathcal{G}$) has at most two T-periodic solutions, namely, the washout solution $t \mapsto u_0^*(t) := (\Sigma_0^*(t), \Sigma_1^*(t), 0)$ and the positive solution $t \mapsto u^*(t) := (\Sigma_0^*(t), \Sigma_1^*(t), y_1^*(t))$, where y_1^* is a component of the T-periodic and positive solution of (5.12) stated in Proposition 5.4. Moreover:*

(i) If the average condition ($\mathcal{A}_1$) is not satisfied, namely,

$$\frac{1}{T}\int_0^T f_1(\Sigma_0^*(t))\,dt \leq D, \tag{5.17}$$

then the washout is the unique T-periodic solution of ($\mathcal{G}$) and is globally attractive.

(ii) If the average condition ($\mathcal{A}_1$) is satisfied, then the positive solution $t \mapsto u^*(t)$ is globally attractive for any set of positive initial conditions.

Proof Firstly, we will focus our attention in the first two equations of ($\mathcal{G}$), which also are an independent subsystem having the form

$$\dot{\eta} = A\eta + \phi(t), \quad \text{where } A = \begin{bmatrix} -D & 0 \\ D & -D \end{bmatrix} \text{ and } \phi(t) = \begin{pmatrix} D[s^0 + be(t)] \\ 0 \end{pmatrix}. \tag{5.18}$$

In order to study the inhomogeneous system (5.18) it is important to note that A is a Hurwitz matrix since its characteristic polynomial is $p(z) = (z + D)^2$ and $D > 0$. This fact has two consequences: Firstly, the corresponding homogeneous linear system $\dot{\eta} = A\eta$ is uniformly asymptotically stable. Secondly, this uniform asymptotic stability property combined with the T-periodicity of $t \mapsto \phi(t)$ and the

Theorem 2.1 from [40] imply that the inhomogeneous system (5.18) has a unique T-periodic solution $t \mapsto \eta^*(t) = (\eta_1^*(t), \eta_2^*(t))$ given by

$$\eta^*(t) := \int_{-\infty}^{t} e^{A(t-s)}\phi(s)\,ds.$$

Note that $t \mapsto \eta_1^*(t)$ is also a T-periodic solution of the scalar equation (5.3). Then by Proposition 5.1, it follows that $\eta_1^*(t) = \Sigma_0^*(t)$ for any t and

$$\lim_{t\to+\infty} \mid \Sigma_0(t) - \Sigma_0^*(t) \mid = 0.$$

Moreover, as $t \mapsto \eta^*(t)$ is a T-periodic solution, it follows that $t \mapsto \eta_2^*(t)$ is also a T-periodic solution of the scalar equation

$$\dot{\Sigma}_1 = -D\Sigma_1 + D\Sigma_0^*(t), \tag{5.19}$$

but, as the above equation has a unique T-periodic solution, it follows that $\eta_2^*(t) = \Sigma_1^*(t)$, where

$$\Sigma_1^*(t) = \int_{-\infty}^{t} De^{-D(t-\tau)}\Sigma_0^*(\tau)\,d\tau,$$

is defined by the recursivity (5.8) for $n = 1$, and any solution $t \mapsto \Sigma_1(t)$ of (5.19) verifies

$$\lim_{t\to+\infty} \mid \Sigma_1(t) - \Sigma_1^*(t) \mid = 0.$$

Note that if $y_1(0) = 0$, then $y_1(t) = 0$ for any $t \geq 0$ and the existence of the washout solution $t \mapsto u_0^*(t) = (\Sigma_0^*(t), \Sigma_1^*(t), 0)$ is verified. The global attractivity of $u_0^*(t)$ follows from the global attractivity of $t \mapsto (\Sigma_0^*(t), \Sigma_1^*(t))$ deduced above combined with Proposition 3.1 from [34, Ch.7].

In order to prove the statement (ii), it is straightforward to see that there exists another T-periodic solution of $(\mathcal{G})$ if and only if the differential equation

$$\dot{y} = f_1(\Sigma_0^*(t) - y)y - Dy \tag{5.20}$$

has a positive T-periodic solution. On the other hand, by using Proposition 3.2 from [34, Ch.7] we know that if the average condition $(\mathcal{A}_1)$ is satisfied then there exists a unique positive T-periodic solution of (5.20), which is globally attractive and the Lemma follows. □

The above result only describe the attractiveness of the T-periodic solution $t \mapsto u^*(t)$ while the next one provides additional properties:

Lemma 5.4 *If the average condition ($\mathcal{A}_1$) is satisfied, then the linearization of the positive solutions of the subsystem ($\mathcal{G}$) around the positive solution $t \mapsto u^*(t) := (\Sigma_0^*(t), \Sigma_1^*(t), y_1^*(t))$ leads to a uniformly asymptotically stable linear system.*

Proof It is easy to see that the linearization of the solutions of ($\mathcal{G}$) around $t \mapsto u^*$ is described by

$$\dot{z} = G_u(t, u^*(t))z, \tag{5.21}$$

where $G_u(t, u^*(t))$ is a positive T-periodic and lower triangular matrix defined as follows

$$\begin{bmatrix} -D & 0 & 0 \\ D & -D & 0 \\ f_1'(\Sigma_0^*(t) - y_1^*(t))y_1^*(t) & 0 & -f_1'(\Sigma_0^*(t) - y_1^*(t))y_1^*(t) + f_1(\Sigma_0^*(t) - y_1^*(t)) - D \end{bmatrix}.$$

As $t \mapsto f_1'(\Sigma_0^*(t) - y_1^*(t))y_1^*(t)$ is continuous and T-periodic, a nice result of F. Batelli and K.J. Palmer stated in [3, Th.1, Th.2] says that the linear system (5.21) is uniformly asymptotically stable if and only if the linear equations:

$$\dot{z}_1 = Az_1 \quad \text{and} \quad \dot{z}_2 = a(t)z_2, \tag{5.22}$$

are uniformly asymptotically stable; where A was defined in (5.18) and

$$a(t) = -[f_1'(\Sigma_0^*(t) - y_1^*(t))y_1^*(t) - f_1(\Sigma_0^*(t) - y_1^*(t)) + D].$$

We claim that the two equations of (5.22) are uniformly asymptotically stable. The first is because A is a Hurwitz matrix, while the second one is a consequence of the fact that a scalar equation $\dot{z} = a(t)z$ is uniformly asymptotically stable if and only if its upper Bohl exponent

$$\beta[a] := \limsup_{t-s \to +\infty} \frac{1}{t-s} \int_s^t a(\tau)\, d\tau$$

is negative [8, Prop.5], [15, Th.3.3.15]. In order to prove its negativeness, it will be useful to note that, as $a(\cdot)$ is T-periodic and continuous, it follows that the Bohl exponent has the simpler description:

$$\beta[a] := \frac{1}{T} \int_0^T a(\tau)\, d\tau.$$

Now, let us recall the identity

$$\int_0^T f_1(\Sigma_0^*(t) - y_1^*(t))\, dt = D.$$

Moreover, as $a(\cdot)$ is T-periodic, its Bohl exponent is given by the average

$$\beta[a] := \frac{1}{T}\int_0^T a(t)\,dt = -\frac{1}{T}\int_0^T f_1'(\Sigma_0^*(t) - y_1^*(t))y_1^*(t)\,dt < 0$$

and the Lemma follows. □

A direct consequence of the above Lemmas is that if the average condition (5.17) is satisfied then $(\mathcal{F})$ is asymptotically T-periodic with limit $\dot{v} = F(u_0^*(t), v)$, namely

$$\begin{cases} \dot{y}_2 = f_1(\Sigma_1^*(t) - y_2 - z_2)y_2 - Dy_2, \\ \dot{z}_2 = f_2(\Sigma_1^*(t) - y_2 - z_2)z_2 - Dz_2. \end{cases} \qquad (\mathcal{F}_0^*)$$

Similarly, if the average condition $(\mathcal{A}_1)$ is satisfied then $(\mathcal{F})$ is asymptotically T-periodic with limit $\dot{v} = F(u^*(t), v)$, namely

$$\begin{cases} \dot{y}_2 = f_1(\Sigma_1^*(t) - y_2 - z_2)y_2 + D[y_1^*(t) - y_2], \\ \dot{z}_2 = f_2(\Sigma_1^*(t) - y_2 - z_2)z_2 - Dz_2. \end{cases} \qquad (\mathcal{F}^*)$$

The study of the planar system $(\mathcal{F}^*)$ plays a fundamental role in the search of a T-periodic solution of the triangular system $(\mathcal{T})$, as shown by the following result, whose proof is straightforward:

Lemma 5.5 *A componentwise positive solution $t \mapsto (u(t), v(t))$ of $(\mathcal{T})$ is T-periodic if and only if $t \mapsto u(t) = u^*(t)$ is the positive T-periodic solution of $(\mathcal{G})$ and $t \mapsto v(t)$ is a T-periodic solution of $(\mathcal{F}^*)$.*

It is important to emphasize that **(H1)** implies that the planar T-periodic systems $(\mathcal{F}_0^*)$ and $(\mathcal{F}^*)$ are *competitive*, that is, the off-diagonal terms of its Jacobian matrix are non positive. Planar competitive systems are ubiquitous in population dynamics and we refer the reader to [35, 36] for details. In particular, the fact that $(\mathcal{F}^*)$ is a competitive system allows a simple proof of the following result:

Proposition 5.5 *For any solution $t \mapsto v(t) = v(t, 0, v_0)$ of $(\mathcal{F}^*)$ passing through the positive initial condition $v_0 = (v_1^0, v_2^0)$ at $t = 0$, there exists $\overline{v} =: (\overline{v}_1, \overline{v}_2)$ such that the corresponding Poincaré map $\mathcal{P}(v_0) = v(T, 0, v_0)$ and its iterations $\mathcal{P}^n(v_0) = v(nT, 0, v_0)$ verify*

$$\lim_{n\to+\infty} \mathcal{P}^n(v_0) = \overline{v} = \mathcal{P}(\overline{v}).$$

Consequently, there exists a T-periodic solution $t \mapsto v^*(t, 0, \overline{v})$ of $(\mathcal{F}^*)$ such that

$$\lim_{t\to+\infty} \mid v(t, 0, v_0) - v^*(t, 0, \overline{v}) \mid = 0.$$

Proof The proof is a direct application of the Theorem 4.2 and Corollary 4.3 from [34, Ch.7]. □

Notice that:

(a) For any positive initial condition v_0 of $(\mathcal{F}^*)$, we can recursively define a sequence of vectors by using the Poincaré map as follows $v_{n+1} = \mathcal{P}(v_n)$, where $v_1 = \mathcal{P}(v_0)$.
(b) The map $v_0 \mapsto \mathcal{P}(v_0) = v(T, 0, v_0)$ is differentiable, this follows from classical results of differentiability of the ODE solutions with respect to the initial conditions [5, 33]. The derivative of $\mathcal{P}$ at v_0 will be denoted by the matrix $\mathcal{P}'(v_0) \in M_2(\mathbb{R})$.
(c) We emphasize that, given two initial conditions v_0 and $\tilde{v}_0$ of $(\mathcal{F}^*)$, the limits of the sequences $\mathcal{P}^n(v_0)$ and $\mathcal{P}^n(\tilde{v}_0)$ in Proposition 5.5 are not necessarily the same. In consequence, Proposition 5.5 ensures the existence of –possibly– several positive T-periodic solutions of the limit system $(\mathcal{F}^*)$ and consequently of the triangular system $(\mathcal{T})$.

In the particular case that the above mentioned Poincaré maps have the same limit $\overline{v}$ regardless the initial condition, we can obtain a more detailed result.

Corollary 5.1 *Assume that the Poincaré map related to initial conditions of* $(\mathcal{F}^*)$ *has a unique positive fixed point* $\overline{v}$ *and that* $\mathcal{P}'(\overline{v})$ *has its two eigenvalues inside the unit circle, then:*

(i) The planar system $(\mathcal{F}^*)$ *has a unique positive* T*-periodic solution* $t \mapsto v^*(t)$*, which is globally attractive for any positive initial condition.*
(ii) The triangular system $(\mathcal{T})$ *has a unique positive* T*-periodic solution* $t \mapsto (u^*(t), v^*(t))$*.*
(iii) The linearization of $(\mathcal{F}^*)$ *around the* T*-periodic solution* $t \mapsto v^*(t)$ *leads to a uniformly asymptotically stable system.*

Proof Let v_0 be a positive initial condition of $(\mathcal{F}^*)$. By using the Proposition 5.5, it follows that $\mathcal{P}^n(v_0)$ converges to $\overline{v}$ when $n \to +\infty$ and $t \mapsto v(t, 0, v_0)$ is attracted by a T-periodic solution $t \mapsto v^*(t, 0, \overline{v})$ and the statement (i) follows. On the other hand, the statement (ii) follows directly from Lemma 5.5.

In order to prove the statement (iii), it is easy to see that the linearization of the solutions of $(\mathcal{F}^*)$ around $t \mapsto v^*(t)$ is described by

$$\dot{z} = F_v(t, u^*(t), v^*(t))z \quad \text{where} \quad F_v(t, u^*(t), v^*(t)) = \begin{bmatrix} a_{11}(t) & a_{12}(t) \\ a_{21}(t) & a_{22}(t) \end{bmatrix} \tag{5.23}$$

with

$$a_{11}(t) = f_1(\Sigma_1^*(t) - y_1^*(t) - z_2^*(t)) - D - f_1'(\Sigma_1^*(t) - y_2^*(t) - z_2^*(t))y_2^*(t),$$
$$a_{12}(t) = -f_1'(\Sigma_1^*(t) - y_2^*(t) - z_2^*(t))y_2^*(t),$$
$$a_{21}(t) = -f_2'(\Sigma_1^*(t) - y_2^*(t) - z_2^*(t))z_2^*(t),$$
$$a_{22}(t) = f_2(\Sigma_1^*(t) - y_1^*(t) - z_2^*(t)) - D - f_2'(\Sigma_1^*(t) - y_2^*(t) - z_2^*(t))z_2^*(t).$$

Notice that the linear system (5.23) is T-periodic and we will denote by $\Phi(t)$ its corresponding fundamental matrix satisfying $\Phi(0) = I$. By Floquet's Theory we know that (5.23) is uniformly asymptotically stable if and only if all the eigenvalues of $\Phi(T)$ are inside the unit circle.

Finally, as pointed out in [34, p.163], the fundamental matrix $\Phi(T)$ verifies the identity

$$\Phi(T) = \mathcal{P}'(\overline{v}),$$

where $\mathcal{P}'(\overline{v})$ is the derivative of the Poincaré map evaluated at the fixed point $\overline{v}$, and the result follows. □

5.3.3 *Coexistence on* $(\mathcal{F})$

Throughout this section, it will be assumed that the average condition $(\mathcal{A}_1)$ is satisfied and the focus will be the coexistence of the species in competition, which motivate us to introduce the following definition

Definition 5.1 ([34, Appendix D]) The system $(\mathcal{F})$ will be called uniformly persistent if there exists $\delta_0 > 0$ such that any solution $t \mapsto (y_2(t), z_2(t))$ verifies

$$\liminf_{t\to+\infty} y_2(t) > \delta_0 \quad \text{and} \quad \liminf_{t\to+\infty} z_2(t) > \delta_0.$$

Lemma 5.6 *Let $t \mapsto (u(t), v(t))$ be a solution of $(\mathcal{T})$ with positive initial condition $v(0) = (y_2(0), z_2(0))$. If the assumptions of Proposition 5.3 are verified, then the less advantaged competitor is uniformly persistent.*

Proof For any positive initial condition of $(\mathcal{T})$ we have that $t \mapsto y_2(t)$ satisfies the differential inequality

$$\dot{y}_2 \geq -Dy_2 + Dy_1(t),$$

since $f_1(\cdot)$ is a nonnegative function. Now, by using classical results of comparison for scalar differential inequalities, see e.g. [18, Lemma 3.4], we can deduce the inequality $0 < w(t) < y_2(t)$ for any $t \geq 0$, where $t \mapsto w(t)$ is any solution of the asymptotically T-periodic equation

$$\dot{w} = -Dw + Dy_1(t)$$

with initial condition $y_2(0) \geq w(0)$. The lemma follows since

$$\lim_{t\to+\infty} w(t) - \int_{-\infty}^{t} De^{-D(t-\tau)} y_1(\tau)\, d\tau = 0.$$

□

An important consequence of Lemma 5.6 is that the chain of chemostats avoids the extinction of the less advantaged competitor provided some mild assumptions as the average condition $(\mathcal{A}_1)$. Nevertheless, the persistence of the more advantaged competitor is not always verified and it will be necessary to find sufficient conditions ensuring it.

Let $t \mapsto (u(t), v(t))$ be a solution of $(\mathcal{T})$ with $u(0) = (\Sigma_0(0), \Sigma_1(0), y_1(0))$ and $v(0) = (y_2(0), z_2(0))$ such that $(u(0), v(0)) \in W$. Notice that if $z_2(0) = 0$, then $z_2(t) = 0$ for any $t \geq 0$, and the first equation of $(\mathcal{F})$ becomes

$$\dot{y}_2 = f_1(\Sigma_1(t) - y_2)y_2 + D[y_1(t) - y_2]. \tag{5.24}$$

Furthermore, as a consequence of Lemma 5.3 we deduce that the above equation is asymptotically T-periodic whose limit equation is

$$\dot{y}_2 = f_1(\Sigma_1^*(t) - y_2)y_2 + D[y_1^*(t) - y_2]. \tag{5.25}$$

Lemma 5.7 *The differential equation (5.25) has at least one T-periodic solution $t \mapsto \overline{y}_2^*(t)$ such that $0 \leq \overline{y}_2^*(t) \leq \Sigma_1^*(t)$. Moreover, if the condition*

$$\min_{t\in[0,T]} \left\{ D + \frac{m_1 k_1 y_1^*(t)}{(\Sigma_1^*(t) + k_1)^2} - \frac{m_1 \Sigma_1^*(t)}{\Sigma_1^*(t) + k_1} \right\} > 0, \tag{5.26}$$

is verified, then the solution $t \mapsto \overline{y}_2^*(t)$ is the unique T-periodic solution of (5.25), and it is globally exponential attractive.

Proof Notice that $(t, u) \mapsto f(t, u)$ defined by

$$f(t, u) = f_1\left(\Sigma_1^*(t) - u\right) y_2 + D\left[y_1^*(t) - u\right]$$

is continuous and of class C^1 with respect to u, since $f_1(u) = m_1 u/(k_1 + u)$. Furthermore, the T-periodic functions $t \mapsto 0(t) \equiv 0$ and $t \mapsto \Sigma_1^*(t)$ are C^1 such that $0(t) \leq \Sigma_1^*(t)$ for $t \in [0, T]$ and, since $f_1(0) = 0$ and $\Sigma_0^* > y_1^*$ it follows that

$$\dot{0}(t) < f\left(t, 0(t)\right), \quad \dot{\Sigma}_1^*(t) > f\left(t, \Sigma_1^*\right), \tag{5.27}$$

therefore $t \mapsto 0(t) \equiv 0$ and $t \mapsto \Sigma_1^*(t)$ are, respectively, lower and upper solutions of (5.25). Then by using monotone iterations, see for instance [1, Th. 3.3 pp 62], it follows that (5.25) has at least one T-periodic solution.

Let us denote by $t \mapsto y_2(t)$ a T-periodic solution of (5.25) and observe that Lemma 5.6 implies that

$$y_2(t) \geq w(t) = \int_{-\infty}^{t} De^{-D(t-\tau)} y_1^*(\tau)\, d\tau \geq \min_{s\in[0,T]} y_1^*(s), \tag{5.28}$$

and it is immediate to see that any other T-periodic solution of (5.25) satisfies the same property.

Let us consider the T-periodic solution $t \mapsto y_2(t)$ and an arbitrary solution $t \mapsto \overline{y}_2(t)$ of (5.24) respectively. Now, we construct the functional

$$V(t) = \frac{1}{2}(y_2(t) - \overline{y}_2(t))^2.$$

Note that the Monod's function f_1 satisfies

$$f_1(u) - f_1(v) = \frac{m_1 k_1 (u - v)}{(u + k_1)(v + k_1)},$$

which allow us to estimate for $\dot{V}(t)$ as follows

$$\begin{aligned}
\dot{V}(t) &= (y_2 - \overline{y}_2)\left\{f_1(\Sigma_1^* - y_2)y_2 + D[y_1^* - y_2] - f_1(\Sigma_1^* - \overline{y}_2)\overline{y}_2 - D[y_1^* - \overline{y}_2]\right\} \\
&= (y_2 - \overline{y}_2)\left\{-D[y_2 - \overline{y}_2] + f_1(\Sigma_1^* - y_2)y_2 - f_1(\Sigma_1^* - \overline{y}_2)\overline{y}_2\right\} \\
&= -2DV(t) + (y_2 - \overline{y}_2)\left\{y_2\left(f_1(\Sigma_1^* - y_2) - f_1(\Sigma_1^* - \overline{y}_2)\right)\right. \\
&\qquad \left. + f_1(\Sigma_1^* - \overline{y}_2)\left[y_2 - \overline{y}_2\right]\right\} \\
&= -2DV(t) + (y_2 - \overline{y}_2)\left\{y_2 \frac{m_1 k_1 (\overline{y}_2 - y_2)}{(\Sigma_1^* - y_2 + k_1)(\Sigma_1^* - \overline{y}_2 + k_1)}\right. \\
&\qquad \left. + \frac{m_1(\Sigma_1^* - \overline{y}_2)(y_2 - \overline{y}_2)}{\Sigma_1^* - \overline{y}_2 + k_1}\right\} \\
&= -2DV(t) + m_1(y_2 - \overline{y}_2)^2\left\{-\frac{k_1 y_2}{(\Sigma_1^* - y_2 + k_1)(\Sigma_1^* - \overline{y}_2 + k_1)} + \frac{(\Sigma_1^* - \overline{y}_2)}{\Sigma_1^* - \overline{y}_2 + k_1}\right\} \\
&< -2DV(t) + m_1(y_2 - \overline{y}_2)^2\left\{-\frac{k_1 y_1^*}{(\Sigma_1^* + k_1)^2} + \frac{\Sigma_1^*}{\Sigma_1^* + k_1}\right\} \\
&< -2\left[D + \frac{m_1 k_1 y_1^*}{(\Sigma_1^* + k_1)^2} - \frac{m_1 \Sigma_1^*}{\Sigma_1^* + k_1}\right] V(t) < 0.
\end{aligned}$$

As consequence of the above inequality and (5.26), it follows that there is a positive constant c such that

$$\ln(V(t)) - \ln(V(0)) < -ct,$$

and the exponential attractivity of any T-periodic solution of (5.25) follows, this way we conclude that there exists a unique positive periodic solution for this equation, which will be denoted by $t \mapsto \bar{y}_2^*(t)$. □

The proof of our next result is based on the nonautonomous semiflows theory introduced by X.Q. Zhao in [42, Ch.3]. Let us recall that Eq. (5.24) depends upon the solutions $t \mapsto u(t)$ of $(\mathcal{G})$ with initial conditions $u(0) = (\Sigma_0(0), \Sigma_1(0), y_1(0))$.

Theorem 5.1 *If the property (5.26) holds, then for any solution $t \mapsto y(t) = y(t, 0, y_0)$ of the Eq. (5.24) it is verified that*

$$\lim_{t\to\infty} | y(t) - \bar{y}_2^*(t) | = 0,$$

for all $0 \leq y_0 < \Sigma_1(0)$, where $t \mapsto \bar{y}_2^*(t)$ is the unique positive T-periodic solution of (5.25).

Proof A direct consequence of the positive invariance of the set W defined in (5.16), is that the solutions of (5.24) and (5.25) are forwardly bounded.

Next, we consider the difference between the Nemytskii operators of (5.24) and (5.25), namely,

$$(t, u) \mapsto \Delta\mathcal{N}(t, u) = u\left(f_1(\Sigma_1(t) - u) - f_1(\Sigma_1^*(t) - u)\right) + D[y_1(t) - y_1^*(t)].$$

Firstly, note that Lemma 5.3 implies that for any $u \in \mathbb{R}$ we have the limit

$$\lim_{t\to+\infty} | \Delta\mathcal{N}(t, u) | = 0. \tag{5.29}$$

Moreover, since Proposition 5.1 ensures that $\Sigma_1(t)$ converges exponentially to $\Sigma_1^*(t)$, we deduce that the above limit is uniform for u in any bounded subset of $\mathbb{R}$.

Secondly, let us define the set

$$\Delta_0 =: \{(t, s) : 0 \leq s \leq t < \infty\}$$

and notice that the solutions of the Eq. (5.24) generate a nonautonomous semiflow $\phi : \Delta_0 \times \mathbb{R} \to \mathbb{R}$ while the solutions of (5.25) generate an T-periodic semiflow $S(t) : \mathbb{R} \to \mathbb{R}$, $t \geq 0$. By using the fact that the limit (5.29) is uniform, it can be easily seen that the Eqs. (5.24) and (5.25) satisfy the assumptions considered by X.Q. Zhao in the Proposition 3.2.2 of [42], which implies that the nonautonomous semiflow $\phi : \Delta_0 \times \mathbb{R} \to \mathbb{R}$ is asymptotic to the T-periodic semiflow $S(t) : \mathbb{R} \to \mathbb{R}$, $t \geq 0$, namely, for any positive integer k, and $r > 0$,

$$\lim_{n\to\infty} | \phi(nT + t, nT, u) - S(t)u | = 0 \quad \text{for any } t \in [0, kT] \text{ and } | u | \leq r.$$

Now, let us define the sequence $\Phi_n(u) := \phi(nT, 0, u)$, for any $u \in [0, \Sigma_1(0)]$ and $n \in \mathbb{N}$. Note that for any $u \in [0, \Sigma_1(0)]$, its positive orbit $\gamma^+(u) = \{\Phi_n(u) : n \geq 0\}$ is bounded in $[0, \Sigma_1(0)]$, and hence its ω-limit set $\omega(u)$ exists.

As the property (5.26) holds, the Lemma 5.7 implies that $u = \bar{y}_2^*(0)$ is a globally asymptotically stable fixed point of the corresponding Poincaré operator $\mathcal{P}$ in $[0, \Sigma_1(0)]$ and $W^s(\bar{y}_2^*(0)) = [0, \Sigma_1(0)]$, where $W^s(\bar{y}_2^*(0))$ is the stable set of $\bar{y}_2^*(0)$ for $\mathcal{P}$.

Finally, for any $u \in [0, \Sigma_1(0)]$, we have that $\omega(u) \cap [0, \Sigma_1(0)] \neq \varnothing$; or equivalently, $\omega(u) \cap W^s(\bar{y}_2^*) \neq \varnothing$. By using Lemma 1.2.3 from [42], we can see that the ω-limit set $\omega(u)$ is internally chain transitive, which also implies that the Theorem 1.2.1 of [42] implies that for any $u \in [0, \Sigma_1(0)]$ it follows that $\omega(u) = \bar{y}_2^*(0)$ and the result follows since $\lim_{n\to\infty} \mid \Phi_n(u) - \bar{y}_2^*(0) \mid = 0$. □

If the average condition $(\mathcal{A}_1)$ is satisfied, a straightforward consequence of the Lemmas 5.3 and 5.6 is that the variables Σ_0, Σ_1, y_1 and y_2 are uniformly persistent. Nevertheless, the uniform persistence of z_2 is not always verified and requires the following complementary average assumption:

$$\frac{1}{T}\int_0^T f_2(\Sigma_1^*(t) - \bar{y}_2^*(t))\, dt > D. \tag{$\mathcal{A}_2$}$$

Next, we present a Lemma that establishes sufficient conditions for uniform persistence of both species in system $(\mathcal{F})$. It is worth mentioning that among the different definitions of persistence from the point of view of applications, uniform persistence is the most stable one under small perturbations [4]. On the other hand, it is immediate to see that the uniform persistence of both species in system $(\mathcal{T})$ depends on the uniform persistence of the microbial species in systems $(\mathcal{G})$ and $(\mathcal{F})$. That is the reason why the following result also establishes sufficient conditions for the uniform persistence of both species in system $(\mathcal{T})$.

Lemma 5.8 *Assume that the inequality (5.26) and the average conditions $(\mathcal{A}_1)$–$(\mathcal{A}_2)$ are verified, then the system $(\mathcal{F})$ is uniformly persistent.*

Proof See Appendix A. □

5.4 Main Results

5.4.1 Preliminaries: Skew-Product Semiflows

The statement of our main result will be made in terms of the skew-product semiflow approach. Firstly, let us write the triangular system in its abbreviated form:

$$\dot{\xi} = q(t, \xi), \tag{$\mathcal{T}$}$$

where $q \in C(W \times \mathbb{R}, \mathbb{R}^5)$ and W was defined in (5.16). Moreover, given $\tau \geq 0$, it will be useful to recall the translation map $q_\tau(t, \xi) := q(t + \tau, \xi)$. Secondly, we will introduce some definitions from [2] and [32, Ch.4] but tailored for the system $(\mathcal{T})$ in order to make this work self-contained:

Definition 5.2 The Hull of q is denoted by $H(q)$ and is composed by all the functions $g \in C(W \times \mathbb{R}, \mathbb{R}^5)$ such that there exists a sequence $\{\tau_n\}_n$ such that $q(t + \tau_n, \xi) \to g(t, \xi)$ uniformly on compact subsets of W.

Definition 5.3 The skew-product semiflow associated with the nonautonomous system $(\mathcal{T})$ is the map $\pi : W \times H(q) \times \mathbb{R}_+ \to W \times H(q)$ described by

$$\pi(x_0, g; t) = (\phi(x_0, g; t), g_t),$$

where $\phi(x_0, g; t)$ is the solution of the initial value problem

$$\begin{cases} \dot{x} = g(t, x) \\ x(0) = x_0. \end{cases} \tag{5.30}$$

As q is T-periodic with respect the variable t, it is straightforward to verify that $g \in H(q)$ if and only if there exists $T_0 \in [0, T)$ such that $q(t + T_0, \xi) = g(t, \xi)$. This characterization of $H(q)$ combined with the fact that the solutions of $(\mathcal{T})$ verify the properties of existence and uniqueness implies that also the solutions of (5.30) satisfy these properties and, in the sense of [32, p.61], we say that q is *regular* or satisfies the *strong uniqueness condition*.

An important consequence of the strong uniqueness condition is given by the Theorem IV.3 from [32], which states that the above defined skew-product semiflow satisfies the continuous semiflow properties. In addition, as the solutions of (5.30) are defined on $\mathbb{R}_+$, it is possible to introduce the ω–limit sets:

Definition 5.4 For any $\xi_0 \in W$, the ω-limit of (ξ_0, q) is defined as the set $\omega(\xi_0, q)$, which is composed by any $(y_0, g) \in W \times H(q)$ such that there exists a divergent sequence $\{\tau_n\}_n$ verifying

$$\lim_{n \to +\infty} \pi(\xi_0, g; \tau_n) = (y_0, g)$$

or equivalently $\lim_{n \to +\infty} \phi(\xi_0, q; \tau_n) = y_0$ and

$$\lim_{n \to +\infty} q_{\tau_n}(t, \xi) = g(t, \xi) \textit{ uniformly on compact subsets of } W.$$

In addition, it is easy to verify that the solutions of (5.30) are uniformly bounded since W is bounded, which allows to say that $\phi(x_0, g; t)$ is *positively compact* in the sense of [32, p.62]. This property leads to the following result

Proposition 5.6 *Let $\xi_0 \in W$ be a positive initial condition of the system $(\mathcal{T})$, then the ω-limit set $\omega(\xi_0, q)$ is nonempty, compact, and invariant, that is, if $(y_0, h) \in \omega(\xi_0, q)$ then $(\phi(y_0, h; t), h_t) \in \omega(\xi_0, q)$ for any $t \geq 0$.*

Proof As q has the regularity property and ϕ is positively compact, the result is a direct consequence of Theorem IV.6 from [32]. □

5.4.2 Main Results

Now, we are able to state our main results.

Theorem 5.2 *Under the assumptions of Proposition 5.3, if the inequality (5.26) and the average conditions $(\mathcal{A}_1)$–$(\mathcal{A}_2)$ are satisfied, then for any positive initial condition $\xi_0 = (u_0, v_0)$ of $(\mathcal{T})$ there exists a T-periodic solution $t \mapsto v^*(t)$ of $(\mathcal{F}^*)$ such that $((u^*(t), v^*(t)), q_t) \in \omega(\xi_0, q)$ for any $t \geq 0$.*

Proof Given a positive initial condition ξ_0 of $(\mathcal{T})$, its corresponding Poincaré map allows to construct the sequence $\{\phi(\xi_0, q; nT)\}_{n\in\mathbb{N}}$. Now, since $t \mapsto \phi(\xi_0, q; t)$ is the solution of $(\mathcal{T})$ passing through $\xi_0 = (u_0, v_0)$ at $t = 0$, we can verify that:

$$\{\phi(\xi_0, q; nT)\}_{n\in\mathbb{N}} := \{(u(nT, 0, u_0), v(nT, 0, \xi_0))\}_{n\in\mathbb{N}}.$$

Because the solutions of $(\mathcal{T})$ are uniformly bounded, we can find a convergent subsequence $v(n_kT, 0, \xi_0) \to \overline{v}_0 := (\bar{v}_1^0, \bar{v}_2^0)$. Moreover, notice that the Lemmas 5.6 and 5.8 imply that $\bar{v}_1^0 > 0$ and $\bar{v}_2^0 > 0$ respectively.

On the other hand, by Lemma 5.3, we know that $u(n_kT, 0, u_0) - u^*(n_kT) \to 0$. But, since $u^*(n_kT) = u^*(0) := u_0^*$, it follows that

$$\{\phi(\xi_0, q; n_kT)\}_{k\in\mathbb{N}} := \{(u(n_kT, 0, u_0), v(n_kT, 0, \xi_0))\}_k \to (u_0^*, \overline{v}_0).$$

By noticing that $q_{n_kT}(t, \xi) = q(t + n_kT, \xi) = q(t, \xi)$ for any t and $k \in \mathbb{N}$, we can deduce that

$$\lim_{k\to+\infty} \pi(\xi_0, q; n_kT) = \lim_{k\to+\infty} (\phi(\xi_0, q; n_kT), q_{n_kT}) = ((u_0^*, \overline{v}_0), q) \in \omega(\xi_0, q).$$

Now, we introduce the notation $\overline{\xi} := (u_0^*, \overline{v}_0)$ and use the invariance result stated by Proposition 5.6 in order to deduce that

$$\pi(\overline{\xi}, q; t) = (\phi(\overline{\xi}, q; t), q_t) \in \omega(\xi_0, q) \quad \text{for any } t \geq 0,$$

which is equivalent to

$$\big((u(t, 0, u_0^*), v(t, 0, \overline{\xi})), q_t\big) \in \omega(\xi_0, q) \quad \text{for any } t \geq 0.$$

Notice that $t \mapsto u(t, 0, u_0^*)$ is the solution of $(\mathcal{G})$ passing through $u^*(0)$ at $t = 0$, then it follows that $u(t, 0, u^*(0)) = u^*(t)$ and the above property can be written as follows

$$\big((u^*(t), v(t, 0, \overline{\xi})), q_t\big) \in \omega(\xi_0, q) \quad \text{for any } t \geq 0. \tag{5.31}$$

It is important to observe that $t \mapsto v(t, 0, \overline{\xi})$ is solution of the system $(\mathcal{F}^*)$.

By using Proposition 5.5, we know that there exists a T-periodic solution $t \mapsto v^*(t) := v^*(t, 0, (u_0^*, \overline{v}))$ of $(\mathcal{F}^*)$ passing through $\bar{v}$ at $t = 0$ such that $v(nT, 0, \bar{\xi}) \to \bar{v}$. Then, by considering $t = nT$ with $n \in \mathbb{N}$ in (5.31), we have that

$$\big((u^*(0), v(nT, 0, \overline{\xi})), q\big) \in \omega(\xi_0, q) \quad \text{for any } n \in \mathbb{N},$$

and, by letting $n \to \infty$, we obtain that $\big((u_0^*, \overline{v}), q\big) \in \omega(\xi_0, q)$.

Finally, by using again the invariance established by Proposition 5.6, we know that $((u^*(t), v^*(t)), q_t) \in \omega(\xi_0, q)$ and the result follows.

□

Theorem 5.2 only ensures the existence of periodic solutions, which is a consequence of Proposition 5.5. Nevertheless, as stated by Remark 5.1, in the case that the T-periodic system $(\mathcal{F}^*)$ has a unique T-periodic solution $t \mapsto v^*(t)$, this implies the existence and uniqueness of a T-periodic solution for $(\mathcal{T})$.

Lemma 5.9 *Under the assumptions of Theorem 5.2 and assuming that the Poincaré map $\mathcal{P}$ associated with the limit system $(\mathcal{F}^*)$ has a unique positive fixed point $\overline{v}$ such that $\mathcal{P}'(\overline{v})$ has its two eigenvalues inside the unit circle, then $(\mathcal{F}^*)$ has a unique T-periodic solution $t \mapsto v^*(t)$ and the linearization of $(\mathcal{T})$ around the T-periodic solution $t \mapsto (u^*(t), v^*(t))$ is uniformly exponentially stable.*

Proof The existence and uniqueness of the solution $t \mapsto v^*(t)$ is a direct consequence of Corollary 5.1. Now, let us consider the error function

$$E(t) = \begin{pmatrix} v(t) - v^*(t) \\ u(t) - u^*(t) \end{pmatrix},$$

where $t \mapsto (u(t), v(t))$ is any positive solution of $(\mathcal{T})$. In consequence, the linearization of $(\mathcal{T})$ around the T-periodic solution $t \mapsto \xi^*(t) = (u^*(t), v^*(t))$ is described by the linear nonautonomous system

$$\dot{E} = L(t)E, \tag{5.32}$$

where $L(t) := Dq(t, \xi^*(t))$ is continuous and satisfies $L(t + T) = L(t)$. It is easy to verify that $L(t)$ is described by the upper triangular matrix

$$L(t) = \begin{bmatrix} F_v(t, u^*(t), v^*(t)) & F_u(t, u^*(t), v^*(t)) \\ 0 & G_u(t, u^*(t)) \end{bmatrix}.$$

By definition, we know that the linearization of $(\mathcal{T})$ around the T-periodic solution $t \mapsto (u^*(t), v^*(t))$ is uniformly exponentially stable if and only if there exists $K > 0$ and $\alpha > 0$ such that any fundamental matrix $\Phi(t)$ of (5.32) verifies

$$\|\Phi(t)\Phi^{-1}(s)\| \leq Ke^{-\alpha(t-s)} \quad \text{for any } t \geq s \geq 0,$$

or equivalently, the linear system (5.32) has an exponential dichotomy on $[0, +\infty)$ with the identity $I \in M_{n_g+n_f}(\mathbb{R})$ as projector.

Because F is a C^1 function, the periodicity of $t \mapsto (u^*(t), v^*(t))$ implies that

$$\sup_{t\geq 0} \mid F_u(t, u^*(t), v^*(t)) \mid < \infty,$$

then, a nice result of F. Batelli and K.J. Palmer stated in the Theorems 1 and 2 from [3] says that the linearization of $(\mathcal{T})$ around the T-periodic solution $t \mapsto (u^*(t), v^*(t))$ is uniformly exponentially stable if and only if, there exists positive constants K_1, K_2, α_1 and α_2 such that

$$\|\Phi_1(t)\Phi_1^{-1}(s)\| \leq K_1e^{-\alpha_1(t-s)} \quad \text{and} \quad \|\Phi_2(t)\Phi_2^{-1}(s)\| \leq K_2e^{-\alpha_2(t-s)},$$

for any $t \geq s \geq 0$, where $\Phi_1(t)$ and $\Phi_2(t)$ are respectively fundamental matrices of the diagonal subsystems

$$\dot{z}_1 = F_v(t, u^*(t), v^*(t))z_1 \quad \text{and} \quad \dot{z}_2 = G_u(t, u^*(t))z_2, \tag{5.33}$$

where $F_v(t, u^*(t), v^*(t))$ was defined in (5.23) and $G_u(t, u^*(t))$ is the matrix introduced in the proof of the Lemma 5.4.

Note that the uniform asymptotic stability of the left linear system of (5.33) follows from the statement (iii) of Corollary 5.1 while the uniform asymptotic stability of the right system is a consequence of the Lemma 5.4.

The proof concludes by noticing that uniform asymptotic stability and uniform exponential stability are equivalent for linear systems, we refer the reader to Theorem 3.3.8 from [15] for details.

□

Theorem 5.3 *Under the assumptions of Theorem 5.2 and assuming that the Poincaré map $\mathcal{P}$ associated with the limit system $(\mathcal{F}^*)$ has a unique positive fixed point $\overline{v}$ such that $\mathcal{P}'(\overline{v})$ has its two eigenvalues inside the unit circle, then $(\mathcal{F}^*)$ has a unique T-periodic solution $t \mapsto v^*(t)$ and the T-periodic solution $t \mapsto (u^*(t), v^*(t))$ is globally attractive for any positive initial condition of the triangular system $(\mathcal{T})$.*

Proof The existence of the T-periodic solution $t \mapsto (u^*(t), v^*(t))$ is a consequence of the previous result. In order to prove its global attractiveness, let us consider again the error function $E(t) = \xi(t) - \xi^*(t)$ where $\xi(t)$ is any solution of $(\mathcal{T})$ with initial condition at ξ_0 and $\xi^*(t) = (u^*(t), v^*(t))$ is the unique T-periodic solution of $(\mathcal{T})$.

Note that $E(t)$ satisfies

$$E' = Dq(t, \xi^*(t))E + h(t, E), \tag{5.34}$$

where h is defined by

$$h(t, E) = q(t, E + \xi^*(t)) - q(t, \xi^*(t)) - Dq(t, \xi^*(t))E$$

and it follows that the origin is an equilibrium of the system (5.34).

Because $E \mapsto q(t, E)$ is differentiable for any fixed t, we have that

$$\lim_{|E| \to 0} \frac{h(t, E)}{| E |} = \lim_{|E| \to 0} \frac{q(t, \xi^*(t) + E) - q(t, \xi^*(t)) - Dq(t, \xi^*(t))E}{| E |} = 0.$$

This can be expressed as follows: given a fixed t, for any $\varepsilon > 0$ there exists $\delta > 0$ such that

$$| h(t, z) | \leq \varepsilon | z | \quad (t, z) \in \mathbb{R}^+ \times B(0, \delta). \tag{5.35}$$

The above property combined with the uniform exponential stability of the linear system $E' = Dq(t, \xi^*(t))E$, ensured by Lemma 5.9, allows us to use Theorem 3.3.41 from [15, p.279] which states that for $\eta \in B(0, r)$ with $r > 0$ sufficiently small, the solution $t \mapsto \varphi(t, t_0, \eta)$ of (5.34) has infinite time life and there exists $M > 0$ and $\theta < 0$ such that

$$\eta \in B(0, r) \Rightarrow | \varphi(t, t_0, \eta) | \leq Me^{-\theta(t - t_0)} | \eta | \quad \text{for all } t \geq t_0. \tag{5.36}$$

Now, if we consider $\eta = E(t_0) = \xi_{t_0} - \xi_{t_0}^*$, then we have that $\varphi(t, t_0, \eta) = \xi(t, t_0, \xi_{t_0}) - \xi^*(t, t_0, \xi_{t_0}^*)$ and the identity (5.36) is equivalent to

$$\xi_{t_0} \in B(\xi_{t_0}^*, r) \Rightarrow | \xi(t, t_0, \xi_{t_0}) - \xi^*(t, t_0, \xi_{t_0}^*) | \leq Me^{-\theta(t - t_0)} | \xi_{t_0} - \xi_{t_0}^* | \quad \text{for all } t \geq t_0.$$

Let $\xi_0 \in W$ be an initial condition of $(\mathcal{T})$, we know by Theorem 5.2 that

$$\left((u^*(t), v^*(t)), q_t\right) \in \omega\left(\xi_0, q\right),$$

that is, there exists a sequence $\{t_k\}_k$ such that $\xi(t_k) - \xi^*(t_k) \to 0$ when $k \to \infty$. Then, for any $r > 0$ small enough, there exists $\hat{k} > 0$ such that $\xi(t_{\hat{k}}) \in B(\xi^*(t_{\hat{k}}), r)$ and the property (5.36) implies that

$$| \xi(t) - \xi^*(t) | \leq Mre^{-\theta(t - t_{\hat{k}})} \quad \text{for any } t > t_{\hat{k}},$$

and the global attractivity of $\xi^*(t) = (u^*(t), v^*(t))$ follows when $t \to +\infty$. □

5.5 Numerical Simulations

To illustrate our results, we will revisit the numerical simulations carried out in [6] and [29], where the limiting nutrient is nitrate, one microbial species is the phytoplankton algae *Dunaniella tertiolecta*[2] while the other one is a theoretical microbial species and the competitive exclusion is verified in an autonomous context. Nevertheless, in order to compare our results, we need to consider a system with the form

$$\begin{cases} \dot{s} = D[s^0 + b\,e(t) - s] - \gamma_1^{-1} f_1(s)z_1 - \gamma_2^{-1} f_2(s)z_2, \\ \dot{z}_1 = z_1[f_1(s) - D], \\ \dot{z}_2 = z_2[f_2(s) - D], \end{cases} \tag{5.37}$$

where the constants γ_1 and γ_2 are called *yield coefficients*. Notice that the changes of variables $x_i = \gamma_i^{-1} z_i$ transforms (5.37) into (5.1).

The uptake functions f_1 and f_2 are of Monod or Michaelis–Menten type as stated by **(H1)**. The input of limiting nutrients is described by the function:

$$s^0(t) = s^0 + 0.01 \sin\left(\frac{2\pi t}{24}\right) \quad \mu\text{mol } L^{-1}$$

and the rest of parameters are presented in Table 5.1.

We point out that the parameters used in [6, 29] and this work are consistent with the intervals presented in the Figure 8 from [37] describing the growth of *Dunaniella tertiolecta* in a chemostat. Before verifying that the hypotheses of Proposition 5.3 are satisfied, we should mention that in the situation presented in this section, unlike that of Sect. 5.1.2, it is species 1 that is the least advantaged, which implies that Proposition 5.3 is valid mutatis mutandis hypotheses (a) and (b). Note that the assumptions of the Proposition 5.3 are verified for $s^0 > \lambda_1$. Indeed, the break-even concentrations satisfy: $\lambda_2 < \lambda_1$ and $m_2 < m_1$. Furthermore, from the values of Table 5.1 it follows that

$$1.16 = \frac{a_2 a_1(\lambda_1 - \lambda_2) + \lambda_1\lambda_2(a_1 - a_2)}{(a_1\lambda_2 - a_2\lambda_1)} > \lambda_1.$$

Consequently, there exists a constant B_1 satisfying $s^0 + 0.01 < B_1$, and $B_1 > 1.16 > \lambda_1$.

The numerical simulations[3] were implemented in R software version 4.0.2 and the libraries DESOLVE, GGPLOT2, RESHAPE2, LATEX2EXP and GRIDEXTRA were used to obtain Fig. 5.2. The simulations were performed considering values of s^0 between 0.03 and 0.15 μmol L^{-1} and in each case condition $(\mathcal{A}_1)$ is satisfied.

[2] https://www.marinespecies.org/aphia.php?p=taxdetails$&$id=178590.

[3] The code is available in https://github.com/Oehninger/ChainChemostat.

Table 5.1 Parameters of the species that compete for the limiting nutrient inside the reactor

Parameter	Value	Unit	Meaning	Source
m_1	1.6	d^{-1}	maximum growth rate species 1	[29]
a_1	0.2	μmolL^{-1}	half-saturation constant species 1	[29]
m_2	1.4	d^{-1}	maximum growth rate species 2	[29]
a_2	0.03	μmolL^{-1}	half-saturation constant species 2	[29]
γ_1	0.35	$\mu\, m^3/\mu$mol	yields species 1	This work
γ_2	0.4	$\mu\, m^3/\mu$mol	yields species 2	This work
D	0.2	d^{-1}	dilution rate	[6]
λ_1	0.02857143	μmol L^{-1}	break-even concentration	[6]
λ_2	0.005	μmol L^{-1}	break-even concentration	[6]

The numerical approximations of the system (5.1) and the last three equations of the system (5.13) are presented in blue and red, respectively. In each graph, species 1 corresponds to the least advantaged while species 2 to the most advantaged. Figure 5.2a–e illustrate our theoretical results since if the washout occurs in the first chemostat of the chain, the most advantaged species will survive; as we increase the average nutrient concentration, the chain will allow both species survive—which does not happen in the system (5.1). Finally, if the average substrate concentration exceeds a threshold, the competitive exclusion is reversed by the chain of the chemostats.

5.6 Robust Uniform Persistence Criterion

5.6.1 Preliminaries

The system ($\mathcal{F}$) has similar structure to the family studied by P.L. Salceanu in [30], which addresses the robust uniform persistence of the systems

$$\begin{cases} \dot{u} = A(t, u, s, \xi)u, \\ \dot{s} = g(t, u, s, \xi), \end{cases} \tag{5.38}$$

where $(u, s) \in \mathbb{R}_+^p \times \mathbb{R}_+^q$ are the state variables, $\xi \in \mathbb{R}^\ell$ is a vector of parameters, $A\colon \mathbb{R}_0^+ \times \mathbb{R}_+^p \times \mathbb{R}_+^q \times \mathbb{R}^\ell \to M_p(\mathbb{R})$ and $g\colon \mathbb{R}_0^+ \times \mathbb{R}_+^p \times \mathbb{R}_+^q \times \mathbb{R}^\ell \to \mathbb{R}_+^q$. The solution of (5.38) passing through $z_0 = (u_0, s_0)$ at time t_0 is denoted by

$$t \mapsto \phi(t + t_0, t_0, z_0, \xi) = (u(t + t_0, t_0, z_0, \xi), s(t + t_0, t_0, z_0, \xi)) .$$

Moreover, the existence, uniqueness, and unbounded forward continuation of the solutions is assumed.

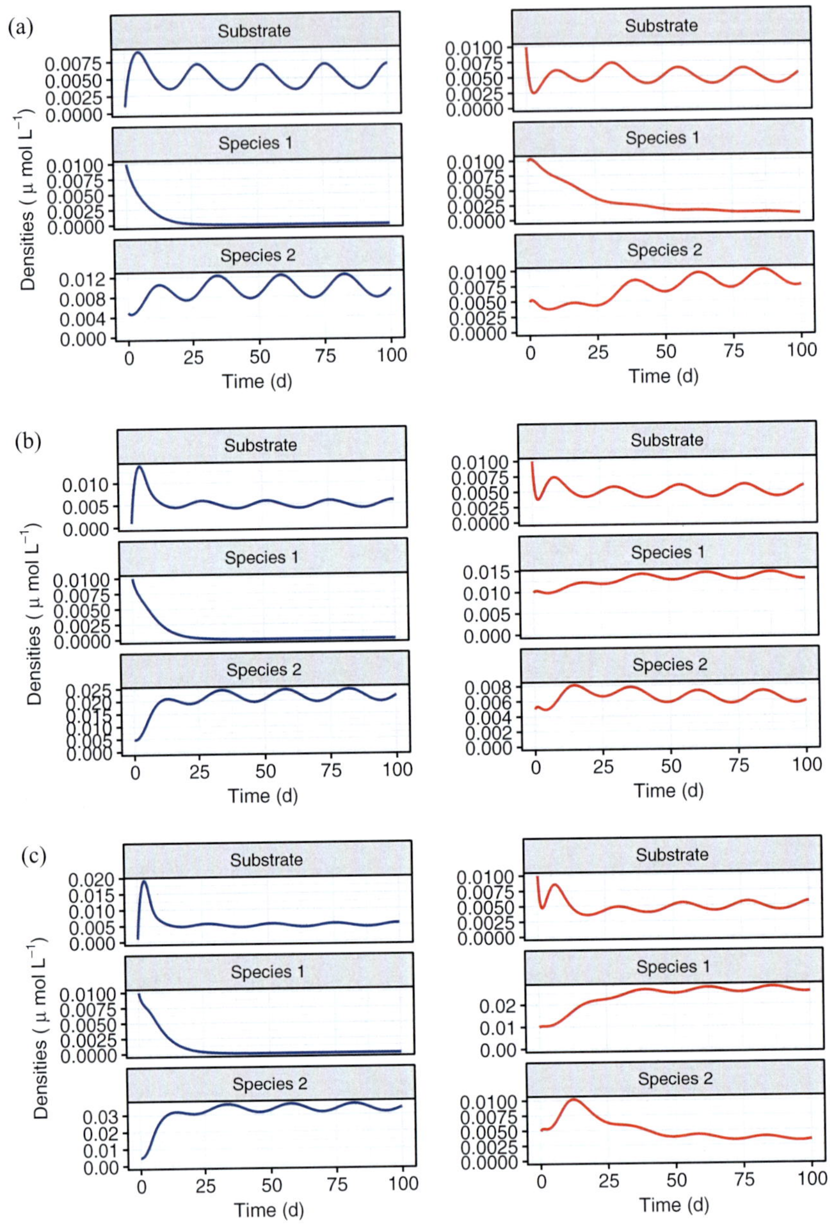

Fig. 5.2 Dynamic of the total biomass and substrate. The initial conditions are: $(s, x_1, x_2) = (0.001, 0.01, 0.005)$ (blue curve) and $(s_1, y_1, s_2, y_2, z_2) = (0.001, 0.01, 0.01, 0.01, 0.005)$ (red curve). (**a**) Dynamics with $s^0 = 0.03$ μmol L^{-1} and $\frac{1}{24}\int_0^{24} f_2(\Sigma_0^*(t))\,dt \approx 0.6970896\ d^{-1}$. (**b**) Dynamics with $s^0 = 0.06$ μmol L^{-1} and $\frac{1}{24}\int_0^{24} f_2(\Sigma_0^*(t))\,dt \approx 0.9332282\ d^{-1}$. (**c**) Dynamics with $s^0 = 0.09$ μmol L^{-1} and $\frac{1}{24}\int_0^{24} f_2(\Sigma_0^*(t))\,dt \approx 1.050639\ d^{-1}$. (**d**) Dynamics with $s^0 = 0.12$ μmol L^{-1} and $\frac{1}{24}\int_0^{24} f_2(\Sigma_0^*(t))\,dt \approx 1.120933\ d^{-1}$. (**e**) Dynamics with $s^0 = 0.15$ μmol L^{-1} and $\frac{1}{24}\int_0^{24} f_2(\Sigma_0^*(t))\,dt \approx 1.167747\ d^{-1}$

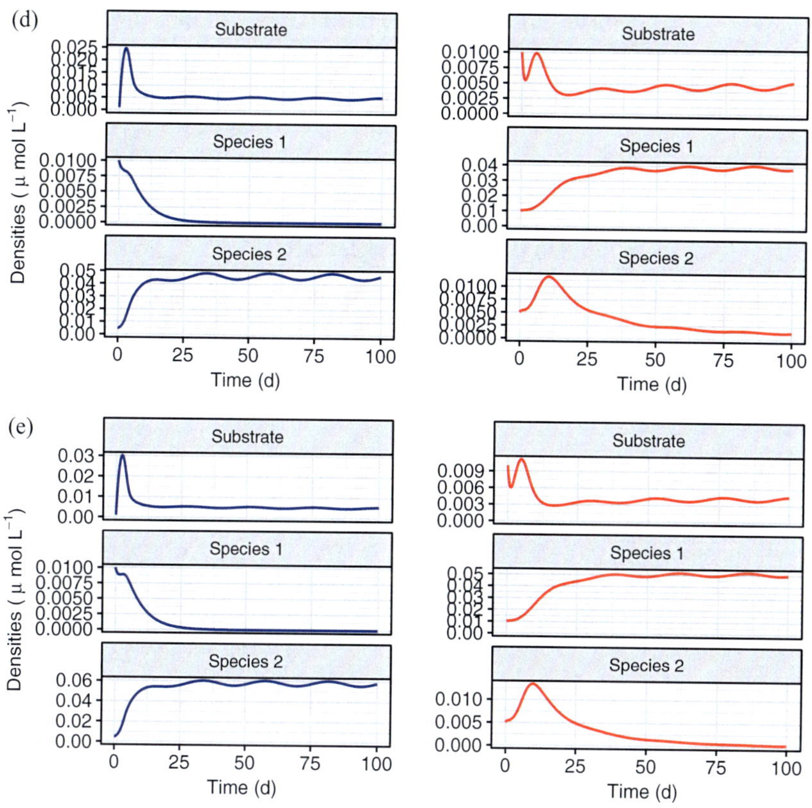

Fig. 5.2 (continued)

An important assumption in [30] is the existence of a positively invariant set $Z \subseteq \mathbb{R}_+^{p+q}$. On the other hand, as $t \mapsto u(t)$ is the biomass concentration of some species, the persistence of u means that the solutions of (5.38) must not converge toward the positively invariant set

$$\Gamma = \left\{(u, s) \in \mathbb{R}_+^p \times \mathbb{R}_+^q : u = 0\right\},$$

and is formally defined as follows:

Definition 5.5 The system (5.38) generating $\phi(t, t_0, z_0, \xi)$ is **robustly uniformly persistent** if there exists $\varepsilon > 0$ and a neighborhood V of the parameter ξ_0 such that

$$\liminf_{t \to +\infty} \mid u(t + t_0, t_0, z, \xi) \mid > \varepsilon \quad \text{for any } t_0 \in \mathbb{R}_0^+, z = (u_0, s_0) \in Z \setminus \Gamma,\ \xi \in V. \tag{5.39}$$

This definition implicitly assumes the positive invariance of $Z \setminus \Gamma$. Moreover, additional assumptions are introduced to address the robust uniform persistence:

(A1) There exists a closed set B absorbing all trajectories corresponding to ξ_0. That is, there exists a time $t(t_0, z) > 0$ such that

$$\phi(t + t_0, t_0, z, \xi_0) \in B \quad \text{for all } t \geq t(t_0, z_0).$$

(A2) Let M and U be defined by

$$M = B \cap \Gamma \quad \text{and} \quad U := \left\{\eta \in \mathbb{R}^q_+ : \mid \eta \mid = 1\right\}.$$

Then, for any $z = (0, s) \in M$ and $\eta \in U$ there exist constants $\ell > 0, c > 1$ and $t_0 \in \mathbb{Z}_0^+$ such that

$$\forall\, t \geq t_0 \, \exists\, \ell(t) \in (0, \ell] \text{ satisfying } \mid P(\ell(t) + t, t, z)\eta \mid > c,$$

where $P(t, t_0, z_0, \xi)$ is the fundamental normalized matrix of the linear system

$$\dot{v} = A(t, \phi(t, t_0, z_0, \xi))v. \tag{5.40}$$

(A3) For all $t_0 \geq 0$ and $z = (0, s) \in M$ there exists a constant $K > 0$ such that $\mid \phi(t + \tau, \tau, z, \xi_0) \mid \leq K$ for any $\tau \geq 0$ and $0 \leq t \leq t_0$.

(A4) There are a neighborhood $\widetilde{V}_B$ of B and a bounded neighborhood $\widetilde{\Xi}$ of ξ_0 such that, for any compact set $\widetilde{M} \subset M$, there exists $K > 0$ such that

$$\begin{aligned}
\|A(t, (0, s), \xi_0)\| &\leq K \\
\mid F(t, (0, s_1), \xi) - F(t, (u_2, s_2), \xi_0) \mid &\leq K(\mid (u_2, s_1 - s_2) \mid + \mid \xi - \xi_0 \mid) \\
\|A(t, (0, s_1), \xi) - A(t, (u_2, s_2), \xi_0)\| &\leq K(\mid (u_2, s_1 - s_2) \mid + \mid \xi - \xi_0 \mid),
\end{aligned}$$

for any $t \geq 0$,$(0, s), (0, s_1) \in \widetilde{M}$,$(u_2, s_2) \in \widetilde{V}_B$ and $\xi \in \widetilde{\Xi}$ where

$$F(t, (u, s), \xi) = (A(t, (u, s), \xi))u, g(t, u, s, \xi))\,.$$

Moreover, for any neighborhood V_B of B, $V_B \subseteq \widetilde{V}_B$, there exists Ξ a neighborhood of ξ_0, $\Xi \subseteq \widetilde{\Xi}$ satisfying the following: $\forall z \in Z$,$s \in \mathbb{R}_0^+$, $\xi \in \Xi$, there exists $t(z, s, \xi) \geq 0$ such that

$$\phi(t + s, s, z, \xi) \in V_B \quad \text{for all } t \geq t(z, s, \xi).$$

(A5) The set $\{(u, s) \in B : \mid s \mid \leq \delta\}$ is bounded for some $\delta > 0$.

Proposition 5.7 (Th.3.5 [30]) *Assume that* **(A1)–(A5)** *hold. Then (5.38) is robustly uniformly persistent.*

The Salceanu's result emulates "in certain sense" the approach of Average Lyapunov functions, which is employed in the autonomous context to determine the repulsiveness of a compact and invariant subset X by studying the dynamics on X and considering some average conditions [16].

5.6.2 *Proof of Lemma 5.8*

It is important to note that the system $(\mathcal{F})$ has a structure similar to (5.38). In this context, let us observe that the set

$$X = \left\{(\Sigma_0, \Sigma_1, y_1, y_2, z_2) \in \mathbb{R}_+^4 \times \mathbb{R}_+ : z_2 = 0\right\}$$

is positively invariant under $(\mathcal{T})$ and the Lemma follows if X is a uniform repeller.

As the bounded set W is positively invariant, it is easy to deduce that the assumptions **(A3)**, **(A4)** and **(A5)** are satisfied. On the other hand, the assumption **(A1)** can be verified by using the boundedness properties of y_1 combined with comparison results for competitive systems.

Finally, in order to verify that the assumption **(A2)** is satisfied, notice that the Eq. (5.40) is equivalent to

$$\dot{w} = \{f_2(\Sigma_1(t) - y_2(t)) - D\}w, \tag{5.41}$$

where $t \mapsto (\Sigma_0(t), \Sigma_1(t), y_1(t), y_2(t), z_2(t))$ is a solution of $(\mathcal{T})$ with initial condition in the set X.

Notice that any solution $t \mapsto w(t)$ of (5.41) passing trough w_0 at $t = t_0$ has the form $w(t) = P(t, t_0)w_0$, where $P(t, s)$ is defined by

$$P(t, s) = e^{\int_s^t \{f_2(\Sigma_1(r) - y_2(r)) - D\}\, dr}.$$

By considering s large enough, Lemma 5.3 and Theorem 5.1 imply that

$$P(t, s) = e^{\int_s^t \{f_2(\Sigma_1^*(r) - \overline{y}_2^*(r) + o(1)) - D\}\, dr},$$

and the average condition $(\mathcal{A}_2)$ implies that

$$P(s + T, s) = e^{\int_s^{s+T} \{f_2(\Sigma_1^*(r) - \overline{y}_2^*(r) + o(1)) - D\}\, dr} = c > 1,$$

then **(A2)** holds and the robust uniform persistence is a consequence of the Proposition 5.7 and the Lemma follows since robust uniform persistence implies uniform persistence.

Acknowledgments This research is partially supported by PROGRAMA REGIONAL MATH-AMSUD MATH2020006. The second author is supported by FONDECYT 1210733 and the third author is supported by FONDECYT 11190457.

References

1. P. Amster, *Topological Methods in the Study of Boundary Value Problems*, vol. 243 (Springer, New York, 2014)
2. Z. Artstein, Limiting equations and stability of nonautonomous ordinary differential equations, in *An appendix in the Stability of Dynamical Systems*, by J.P. LaSalle. Regional Conference Series in Applied Mathematics, vol. 25 (SIAM Publications, New York, 1976)
3. F. Battelli, K.J. Palmer, Criteria for exponential dichotomy for triangular systems, J. Math. Anal. Appl. **428**, 525–543 (2015)
4. G. Butler, H.I. Freedman, P. Waltman, Uniformly persistent systems, in *Proceedings of the American Mathematical Society*, vol. 96 (1986), pp. 425–430
5. E. Coddington, N. Levinson, *Theory of Ordinary Differential Equations* (Mc Graw–Hill, New York, 1955)
6. G. Contreras, *Mathematical Model of Competition in a Chain of Chemostats: Some Results About Coexistence and Exclusion*, Master Thesis (Universidad de Chile, Spanish, 2013)
7. S. Dikshitulu, B. Baltzis, G. Lewandowski, S. Pavlou, Competition between two microbial populations in a sequencing fed-batch reactor: theory, experimental verification, and implications for waste treatment applications. Biotechnol. Bioeng. **42**, 643–656 (2004)
8. T.S. Doan, K.J. Palmer, M. Rasmussen, The Bohl spectrum for linear nonautonomous differential equations. J. Dyn. Diff. Eqs. **29**, 1459–1485 (2017)
9. G. Fan, H.L. Smith, H.R. Thieme, Competition in the chemostat with time-dependent differential removal rates. Vietnam. J. Math. **45**, 153–178 (2017)
10. P. Gajardo, F. Mazenc, H. Ramirez, Competitive exclusion principle in a model of chemostat with delays. Dynam. Contin. Dis. Ser. A **16**, 253–272 (2009)
11. J. Hale, A.S. Somolinos, Competition for a fluctuating nutrient. J. Math. Biol. **18**, 255–280 (1983)
12. S. Hansen, S. Hubell, Single nutrient microbial competition: agreement between experimental and forecast outcomes. Science **207**, 1491–1493 (1980)
13. G. Hardin, Competitive exclusion principle. Science **131**, 1292–1297 (1960)
14. J. Harmand, C. Lobry, A. Rapaport, T. Sari, *The Chemostat: Mathematical Theory of Microorganism Cultures*, vol. 1 (ISTE–Wiley, New York, 2017)
15. D. Hinrichsen, A.J. Pritchard, *Mathematical Systems Theory I* (Springer, Berlin, 2005)
16. J. Hofbauer, A unified approach to persistence. Acta Appl. Math. **14**, 11–22 (1989)
17. J.L. Jost, J.F. Drake, A.G. Frederickson, H.M Tsuchiya, Interactions of *Tetrahymena pyriformis, Escherichia coli, Azotobacter vinelandii* and glucose in a minimal medium. J. Bacteriol. **113**, 834–840 (1973)
18. H. Khalil, *Nonlinear Systems*. (Prentice Hall, Upper Saddle River, 1996)
19. B. Li, Global asymptotic behavior of the chemostat: general response functions and different removal rates. SIAM J. Appl. Math. **59**, 411–422 (1999)
20. B. Li, G.S.K. Wolkowicz, Y. Kuang, Global asymptotic behavior of a chemostat model with two perfectly complementary resources and distributed delay. SIAM J. Appl. Math. **60**, 2058–2086 (2000)
21. S. Liu, X. Wang, L. Wang, H. Song, Competitive exclusion in delayed chemostat models with differential removal rates. SIAM J. Appl. Math. **74**, 634–648 (2014)
22. F. Mazenc, S.I. Niculescu, G. Robledo, Stability analysis of mathematical model of competition in a chain of chemostats in series with delay. Appl. Math. Model. **76**, 311–329 (2019)
23. J. Monod, *Recherches sur la croissance des cultures bactériennes* (Hermann, Paris, 1942)
24. J. Monod, The growth of bacterial cultures. Ann. Rev. Microbiol. **3**, 371–394 (1949)
25. J. Monod, La technique de culture continue, théorie et applications. Ann. l'Inst. Pasteur **79**, 390–410 (1950)
26. A. Novick, L. Slizard, Description of the chemostat. Science **112**, 715–716 (1950)

27. E. Postma, A. Kuiper, W. Tomasouw, W. Scheffers, J. Van Dijken, Competition for glucose between the yeasts Saccharomyces cerevisiae and Candida utilis. Appl. Environ. Microbiol. **55**, 3124–3220 (1989)
28. A. Rapaport, M. Veruete, A new proof of the competitive exclusion principle in the chemostat. Discrete Contin. Dyn. Syst. Ser. B **24**, 3755–3764 (2019)
29. G. Robledo, F. Grognard, J.L. Gouzé, Global stability for a model of competition in the chemostat with microbial inputs. Nonlinear Anal. Real World Appl. **13**, 582–598 (2012)
30. P.L. Salceanu, Robust uniform persistence in discrete and continuous nonautonomous systems. J. Math. Anal. Appl. **398**, 487–500 (2013)
31. T. Sari, Competitive exclusion for chemostat equations with variable yields. Acta Appl. Math. **123**, 201–219 (2013)
32. G. Sell, *Topological Dynamics and Ordinary Differential Equations* (Van Nostrand–Reinhold, London, 1971)
33. T.C. Sideris, *Ordinary Differential Equations and Dynamical Systems* (Atlantis Press, Paris, 2013)
34. H.L. Smith, P. Waltman, *The Theory of the Chemostat. Dynamics of Microbial Competition* (Cambridge University, Cambridge, 1995)
35. H.L. Smith, Dynamics of competition, in *Mathematics Inspired by Biology Lectures Given at the 1st Session of the Centro Internazionale Matematico Estivo (C.I.M.E.) Held in Martina Franca, Italy, June 13—20, 1997*, ed. by V. Capasso, pp. 191–240. Lecture Notes in Mathematics, vol. 1714 (Springer, Berlin, 1999)
36. H.L. Smith, *Monotone Dynamical Systems: An Introduction to the Theory of Competitive and Cooperative Systems, Mathematical Surveys and Monographs*, vol. 41 (Amer. Math. Soc., Providence, 1995)
37. I. Vatcheva, H. de Jong, O. Bernard, N.J.I. Mars, Experiment selection for the discrimination of semi-quantitative models of dynamical systems. Artif. Intell. **170**, 472–506 (2006)
38. G.S.K. Wolkowicz, X.Q. Zhao, N-species competition in a periodic chemostat. Differ. Integr. Equations **11**, 465–491 (1998)
39. C. Xu, S. Yuan, T. Zhang, Competitive exclusion in a general multi-species chemostat model with stochastic perturbations. Bull. Math. Biol. **83**, Paper No. 4, 17 pp. (2021)
40. C. Zhang, Pseudo almost periodic solutions of some differential equations. J. Math. Anal. Appl. **181**, 62–76 (1994)
41. Q. Zhang, D. Jiang, Competitive exclusion in a stochastic chemostat model with Holling type II functional response. J. Math. Chem. **54**, 777–791 (2016)
42. X.Q. Zhao, *Dynamical Systems in Population Biology*. CMS Books in Mathematics (Canadian Mathematical Society—Springer, New York, 2003)

Chapter 6
Nontrivial Solutions of a Parameter-Dependent Heat-Flow Problem with Deviated Arguments

Alessandro Calamai and Gennaro Infante

MSC 2020 Primary 34B08, Secondary 34B10, 34K10, 47H10

6.1 Introduction

In this chapter we investigate the existence of *nontrivial* solutions of the second-order parameter-dependent differential equation

$$u''(t) + \lambda f(t, u(t), u(\sigma(t))) = 0, \ t \in [0, 1], \tag{6.1}$$

with initial conditions

$$u(t) = \omega(t), \ t \in [-r, 0], \tag{6.2}$$

and the functional boundary condition (BC)

$$\beta u'(1) + u(\eta) = \lambda B[u], \tag{6.3}$$

where $\beta > 0$, $\eta \in (0, 1)$, and $0 < \beta + \eta < 1$, λ is a parameter, σ, ω, f are suitable continuous functions, and B is a suitable functional.

A. Calamai (✉)
Dipartimento di Ingegneria Civile, Edile e Architettura, Università Politecnica delle Marche Via Brecce Bianche, Ancona, Italy
e-mail: a.calamai@univpm.it

G. Infante
Dipartimento di Matematica e Informatica, Università della Calabria, Cosenza, Italy
e-mail: gennaro.infante@unical.it

P. Amster, P. Benevieri (eds.), *Topological Methods for Delay and Ordinary Differential Equations*, Advances in Mechanics and Mathematics 51,
https://doi.org/10.1007/978-3-031-61337-1_6

The motivation for studying this problem is that it arises in heat-flow problems. To illustrate this in a simple situation, let us consider the following special case of a problem with reflection of the argument:

$$\begin{cases} u''(t) + \lambda f(t, u(t), u(-t + \frac{1}{2})) = 0, & t \in [0, 1], \\ u(t) = \omega(t), & t \in [-\frac{1}{2}, 0], \\ \beta u'(1) + u(\frac{1}{4}) = 0. \end{cases} \tag{6.4}$$

The BVP (6.4) describes the steady states of temperature u of a heated bar of length $3/2$. The left end of the bar is kept at a prescribed temperature ω, and a controller, placed at the right end of the bar, is reacting to a sensor placed in the point $t = \frac{1}{4}$. The presence of the term depending on the reflection of the argument illustrates the influence of the left-hand side of the bar on the right-hand side; a physical problem of this kind occurs in the case of a light bulb, see [1] for a detailed description. For simplicity here we have taken $B \equiv 0$ in the BCs; a nonzero term B can be used to model complicated controller, possibly with nonlinear response. There exists a wide literature on similar kinds of heat-flow problems; for brevity we mention the papers [2, 5, 7, 9, 15, 16, 21, 23–26] in the case of *linear* controllers and [12, 17–19, 21, 22] for *nonlinear* controllers. Note that different choices of the function σ lead to different situations; for example a spatial delay can be modelled by $\sigma(t) = t-\tau$, where $\tau \leq r$; this has been done, for example, in the context of delay equations in [2]. More general deviated arguments occurring in heat-flow problems have been studied in [1, 8].

In order to discuss the solvability of the BVP (6.1)–(6.2)–(6.3), we adopt a topological approach, based on a recent variant [3] in affine cones of the celebrated Birkhoff–Kellogg theorem. We provide an example where we illustrate the constants that occur in our theory.

Topological approaches in affine cones have been exploited recently in [2–4, 6]. The setting of affine cones seems to be helpful when dealing with equations with delay effects. Here we prove the existence of nontrivial solutions (u, λ) of the BVP (6.1)–(6.2)–(6.3), by means of an associated *perturbed* Hammerstein integral equation set in a suitable translate of a cone of functions that are allowed to *change sign*. We mention that another variant of the Birkhoff–Kellogg theorem, due to Krasnosel'skiĭ and Ladyženskiĭ [20], has been used in [10] for a cone of sign-changing functions with vertex in the origin under different sets of boundary conditions and without the presence of deviated arguments.

Our results are new and complement the previous literature. In particular we use a different topological tool w.r.t. [1, 2], where the fixed point index is used directly. Furthermore here we study a BVP that is different from the one in [2], due to the presence of the deviated arguments and the nonlinearity within the BCs, and from the one in [1], due to the presence of the datum ω.

6.2 Solutions of Perturbed Integral Equations in Affine Cones

We recall some useful notation. Let $(X, \| \ \|)$ be a real Banach space. A *cone* K of X is a closed set with $K + K \subset K$, $\mu K \subset K$ for all $\mu \geq 0$ and $K \cap (-K) = \{0\}$. For $y \in X$, the *translate* of the cone K is defined as

$$K_y := y + K = \{y + x : x \in K\}.$$

Given a bounded and open (in the relative topology) subset Ω of K_y, we denote by $\overline{\Omega}$ and $\partial\Omega$ the closure and the boundary of Ω relative to K_y. Given an open bounded subset D of X, we denote $D_{K_y} = D \cap K_y$, an open subset of K_y.

We can now state a Birkhoff–Kellogg-type result, which provides the existence of nontrivial solutions of parameter-dependent functional equations in cone translates.

Theorem 6.1 ([3], Corollary 2.4) *Let $(X, \| \ \|)$ be a real Banach space, $K \subset X$ be a cone, and $D \subset X$ be an open bounded set with $y \in D_{K_y}$ and $\overline{D}_{K_y} \neq K_y$. Assume that $\mathcal{F} : \overline{D}_{K_y} \to K$ is a compact map and assume that*

$$\inf_{x \in \partial D_{K_y}} \|\mathcal{F}(x)\| > 0.$$

Then there exist $x^* \in \partial D_{K_y}$ and $\lambda^* \in (0, +\infty)$ such that $x^* = y + \lambda^* \mathcal{F}(x^*)$.

In order to apply Theorem 6.1, we make some assumptions on the following *perturbed* Hammerstein integral equation:

$$\begin{aligned} u(t) &= \psi(t) + \lambda\Big(\int_0^1 k(t,s)g(s)f(s,u(s),u(\sigma(s))\,ds + \gamma(t)B[u]\Big) \\ &=: \psi(t) + \lambda\mathcal{F}u(t), \ \ t \in [-r, 1], \end{aligned} \tag{6.5}$$

where B is a suitable (possibly nonlinear) functional in the space $C([-r, 1], \mathbb{R})$, endowed of the usual supremum norm, $\|u\|_{[-r,1]}$. More in general, given a compact interval $I \subset \mathbb{R}$, we denote by $C(I, \mathbb{R})$ the Banach space of the continuous functions defined on I with the usual norm, $\|u\|_I$. We assume the following conditions:

(C_1) The function $\psi : [-r, 1] \to \mathbb{R}$ is continuous.
(C_2) The kernel $k : [-r, 1] \times [0, 1] \to \mathbb{R}$ is measurable, verifies $k(t, s) = 0$ for all $t \in [-r, 0]$ and almost every (a. e.) $s \in [0, 1]$, and for every $\bar{t} \in [0, 1]$ we have

$$\lim_{t \to \bar{t}} |k(t, s) - k(\bar{t}, s)| = 0 \ \text{ for a. e. } s \in [0, 1].$$

(C_3) There exist a subinterval $[a, b] \subseteq [0, 1]$, a measurable function Φ with $\Phi \geq 0$ a. e., and a constant $c_1 = c_1(a, b) \in (0, 1]$ such that

$$|k(t, s)| \leq \Phi(s) \text{ for all } t \in [0, 1] \text{ and a. e. } s \in [0, 1],$$
$$k(t, s) \geq c_1 \Phi(s) \text{ for all } t \in [a, b] \text{ and a. e. } s \in [0, 1].$$

(C_4) The function $g : [0, 1] \to \mathbb{R}$ is measurable, $g(t) \geq 0$ a. e. $t \in [0, 1]$, and satisfies that $g\,\Phi \in L^1[0, 1]$ and $\int_a^b \Phi(s)g(s)\,ds > 0$.

(C_5) $f : [0, 1] \times \mathbb{R} \times \mathbb{R} \to [0, \infty)$ satisfies some Carathéodory-type conditions; namely, $f(\cdot, u, v)$ is measurable for each fixed u and v in $\mathbb{R}$, $f(t, \cdot, \cdot)$ is continuous for a. e. $t \in [0, 1]$, and for each $R > 0$, there exists $\phi_R \in L^\infty[0, 1]$ such that

$$f(t, u, v) \leq \phi_R(t) \;\text{ for all }\; (u, v) \in [-R, R]\times[-R, R], \;\text{ and a. e. }\; t \in [0, 1].$$

(C_6) The function $\sigma : [0, 1] \to [-r, 1]$ is continuous.

(C_7) The function $\gamma : [-r, 1] \to \mathbb{R}$ is continuous, $\gamma \not\equiv 0$, and such that $\gamma(t) = 0$ for all $t \in [-r, 0]$; moreover there exists $c_2 \in (0, 1]$ such that $\gamma(t) \geq c_2\|\gamma\|_{[0,1]}$ for all $t \in [a, b]$.

In the Banach space $C([-r, 1], \mathbb{R})$ we define the cone

$$K_0 = \{u \in C([-r, 1], \mathbb{R}) : u(t) = 0 \text{ for all } t \in [-r, 0], \min_{t\in[a,b]} u(t) \geq c\|u\|_{[0,1]}\},$$

where $c = \min\{c_1, c_2\}$. Note that $K_0 \neq \{0\}$ since $\gamma \in K_0$ and, furthermore, that the functions in K_0 are nonnegative in the subset $[a, b]$ and may *change sign* in $[0, 1]$. The cone K_0 has been essentially introduced [2], as a modification of a cone introduced in [13].

We consider the following translate of the cone K_0,

$$K_\psi = \psi + K_0 = \{\psi + u : u \in K_0\},$$

with the subset

$$K_{0,\rho} := \{u \in K_0 : \|u\|_{[0,1]} < \rho\}$$

and the corresponding translate

$$K_{\psi,\rho} := \psi + K_{0,\rho}.$$

Note that $\partial K_{\psi,\rho} = \psi + \partial K_{0,\rho}$ and that $u \in K_\psi$ means that $u = \psi + v$ with $v \in K_0$ and, therefore, we have

$$\|u\|_{[-r,1]} = \max\{\|\psi\|_{[-r,0]}, \|\psi + v\|_{[0,1]}\}.$$

We can now state our existence result.

Theorem 6.2 *Let $\rho \in (0, +\infty)$ and assume the following conditions hold:*

(a) *There exist $\underline{\delta}_\rho \in C([0,1], \mathbb{R}_+)$ such that*

$$f(t,u,v) \geq \underline{\delta}_\rho(t), \text{ for every } (t,u,v) \in [a,b] \times \mathbb{R} \times \mathbb{R} \text{ with}$$
$$\max\{|u|, |v|\} \leq \rho + \|\psi\|_{[-r,1]}.$$

(b) *$B : \overline{K}_{\psi,\rho} \to \mathbb{R}_+$ is continuous and bounded; in particular let $\underline{\eta}_\rho \in [0, +\infty)$ be such that*

$$B[u] \geq \underline{\eta}_\rho, \text{ for every } u \in \partial K_{\psi,\rho}.$$

(c) *The inequality*

$$\sup_{t\in[a,b]} \left\{ \gamma(t)\underline{\eta}_\rho + \int_a^b k(t,s)g(s)\underline{\delta}_\rho(s)\, ds \right\} > 0 \tag{6.6}$$

holds.

Then there exist $\lambda_\rho \in (0, +\infty)$ and $u_\rho \in \partial K_{\psi,\rho}$ that satisfy the integral equation (6.5).

Proof We firstly show that the operator $\mathcal{F}$ maps $\overline{K}_{\psi,\rho}$ into K_0 and is compact. Take $u \in \overline{K}_{\psi,R}$; we have to prove that $\mathcal{F}u \in K_0$. First of all observe that our assumptions imply that $\mathcal{F}u$ is continuous on $[-r, 1]$ and that $\mathcal{F}u(t) = 0$ for $t \in [-r, 0]$. Now, for every $t \in [0, 1]$ we have

$$\begin{aligned}|\mathcal{F}u(t)| &\leq \int_0^1 |k(t,s)|g(s)f(s,u(s),u(\sigma(s)))\, ds + |\gamma(t)|B[u] \\ &\leq \int_0^1 \Phi(s)g(s)f(s,u(s),u(\sigma(s)))\, ds + \|\gamma\|_{[0,1]}B[u],\end{aligned}$$

and moreover, for $t \in [a, b]$,

$$\mathcal{F}u(t) \geq c_1 \int_0^1 \Phi(s)g(s)f(s,u(s),u(\sigma(s)))\, ds + c_2\|\gamma\|_{[0,1]}B[u] \geq c\|\mathcal{F}u\|_{[0,1]}. \tag{6.7}$$

Taking the minimum for $t \in [a, b]$ in (6.7) yields $\mathcal{F}u \in K_0$, as desired. The compactness of $\mathcal{F}$ follows in a similar way as in the proof of Theorem 3.2 of [2], where a linear functional was considered in place of B. Note that here the functional B (possibly nonlinear) is continuous and bounded.

Now, take $u \in \partial K_{\psi,\rho}$. Then we have, for $t \in [a, b]$,

$$\begin{aligned}\mathcal{F}u(t) &= \Big(\int_0^1 k(t,s)g(s)f(s,u(s),u(\sigma(s)))\,ds + \gamma(t)B[u]\Big)\\ &\geq \Big(\int_a^b k(t,s)g(s)f(s,u(s),u(\sigma(s)))\,ds + \gamma(t)B[u]\Big)\\ &\geq \gamma(t)\underline{\eta}_\rho + \int_a^b k(t,s)g(s)\underline{\delta}_\rho(s)\,ds.\end{aligned}$$

Therefore we have

$$\|\mathcal{F}u\|_{[-r,1]} \geq \|\mathcal{F}u\|_{[a,b]} \geq \sup_{t\in[a,b]}\Big\{\gamma(t)\underline{\eta}_\rho + \int_a^b k(t,s)g(s)\underline{\delta}_\rho(s)\,ds\Big\}. \tag{6.8}$$

Note that the RHS of (6.8) does not depend on the particular u chosen. Hence,

$$\inf_{u\in\partial K_{\psi,\rho}} \|\mathcal{F}u\|_{[-r,1]} \geq \sup_{t\in[a,b]}\Big\{\gamma(t)\underline{\eta}_\rho + \int_a^b k(t,s)g(s)\underline{\delta}_\rho(s)\,ds\Big\} > 0,$$

and the result follows by Theorem 6.1. □

6.3 Nontrivial Solutions of the BVP

We turn our attention back to the BVP

$$\begin{cases} u''(t) + \lambda f(t,u(t),u(\sigma(t))) = 0, & t \in [0,1],\\ u(t) = \omega(t), & t \in [-r,0],\\ \beta u'(1) + u(\eta) = \lambda B[u]. \end{cases} \tag{6.9}$$

In order to apply the previous theory to the BVP (6.9), we proceed by means of a superposition principle as in Section 3 of [4], in the spirit of [2, 11, 14].

We begin by considering the BVP

$$\begin{cases} u''(t) + y(t) = 0, & t \in [0,1],\\ u(t) = 0,\ \beta u'(1) + u(\eta) = 0, \end{cases}$$

which has the has the unique solution

$$u(t) = \int_0^1 \hat{k}(t,s)y(s)ds,$$

where the Green's function (see for example [14]) is given by

$$\hat{k}(t,s) = \frac{\beta t}{\beta+\eta} + \frac{t}{\beta+\eta}(\eta - s)H(\eta - s) - (t-s)H(t-s),$$

where

$$H(\tau) = \begin{cases} 1, & \tau \geq 0, \\ 0, & \tau < 0, \end{cases}$$

and thus we take

$$k(t,s) = \hat{k}(t,s)H(t). \tag{6.10}$$

With the choice of $[a,b] \subset (0, \beta+\eta) \subset (0,1)$, the hypotheses (C_2)–(C_3) are satisfied (see [2]) when

$$\Phi(s) = \begin{cases} s, & \text{for } \beta+\eta \geq \frac{1}{2}, \\ \left[\dfrac{1-(\beta+\eta)}{\beta+\eta}\right] s, & \text{for } \beta+\eta < \frac{1}{2}, \end{cases}$$

and

$$c_1 = \begin{cases} \min\left\{\dfrac{a\beta}{\beta+\eta}, \dfrac{\beta+\eta-b}{\beta+\eta}\right\}, & \text{for } \beta+\eta \geq \frac{1}{2}, \\ \min\left\{\dfrac{a\beta}{1-(\beta+\eta)}, \dfrac{\beta+\eta-b}{1-(\beta+\eta)}\right\}, & \text{for } \beta+\eta < \frac{1}{2}. \end{cases} \tag{6.11}$$

Note that also (C_4) holds since $\int_a^b \Phi(s)\,ds > 0$.

Now observe that the function $\hat{\gamma}(t) = \dfrac{t}{\beta+\eta}$ satisfies the BVP

$$\begin{cases} u''(t) = 0, \ t \in [0,1], \\ u(0) = 0, \ \beta u'(1) + u(\eta) = 1. \end{cases}$$

Thus we may take

$$\gamma(t) = \hat{\gamma}(t)H(t), \tag{6.12}$$

and note that (C_7) is satisfied with $c_2 = a$ (see [2]). Note also that $c_2 \geq c_1$; hence we take $c = c_1$.

Finally note that $\hat{\varphi}(t) = \dfrac{(\beta+\eta)-t}{\beta+\eta}$ solves the BVP

$$\begin{cases} u''(t) = 0, \; t \in [0,1], \\ u(0) = 1, \; \beta u'(1) + u(\eta) = 0 \end{cases}.$$

Thus, define

$$\psi(t) = \begin{cases} \omega(t), & t \leq 0, \\ \hat{\varphi}(t)\omega(0), & t > 0, \end{cases} \tag{6.13}$$

and observe that, by construction, ψ is continuous on $[-r, 1]$, since ω is assumed to be continuous on $[-r, 0]$. Note that ψ, the vertex of the affine cone that we utilize, is built from the initial datum ω, for which we do not require homogeneity in 0, as in [4].

Summing up, we can consider the integral equation

$$u(t) = \psi(t) + \lambda\Big(\int_0^1 k(t,s) f(s, u(s), u(\sigma(s)))\, ds + \gamma(t) B[u]\Big), \quad t \in [-r, 1], \tag{6.14}$$

which is of the form (6.5) where $g \equiv 1$, k is as in (6.10), γ is as in (6.12), and ψ is as in (6.13).

Definition 6.1 By a solution of the BVP (6.9) we mean a solution $u \in C([-r, 1], \mathbb{R})$ of the integral equation (6.14).

With the above ingredients, we can state the following existence result.

Theorem 6.3 *Let $f : [0,1] \times \mathbb{R} \times \mathbb{R} \to [0, \infty)$ and $\sigma : [0,1] \to [-r, 1]$ be continuous. Let $[a, b] \subset (0, \beta + \eta) \subset (0, 1)$, and let $c = c_1$ as in* (6.11). *Let $\rho \in (0, +\infty)$ and assume that conditions (a), (b), and (c) of Theorem 6.2 hold. Then there exist $\lambda_\rho \in (0, +\infty)$ and $u_\rho \in \partial K_{\psi,\rho}$ that satisfy the BVP* (6.9).

In the next illustrating example we consider a specific BVP of type (6.9), with the choice $r = 1/2$ and $\sigma(s) = -s$, namely an equation with reflection of the argument.

Example 6.4 We consider the BVP

$$\begin{cases} u''(t) + \lambda t e^{u(t)+2u(-t+\frac{1}{2})} = 0, & t \in [0,1], \\ u(t) = \sqrt{1+t}, & t \in [-\frac{1}{2}, 0], \\ \frac{1}{4}u'(1) + u(\frac{1}{4}) = \lambda \int_{-1}^{1} t^2 (u(t))^2\, dt. \end{cases} \tag{6.15}$$

Thus the function ψ is given by

$$\psi(t) = \begin{cases} \sqrt{1+t}, & -\frac{1}{2} \leq t \leq 0, \\ 1 - 2t, & 0 < t \leq 1. \end{cases}$$

Now choose $\rho \in (0, +\infty)$ and note that $\|\psi\|_{[-\frac{1}{2},1]} = 1$. We may take

$$[a, b] = \left[\frac{1}{8}, \frac{1}{4}\right], \ \underline{\eta}_\rho(t) = 0, \ \underline{\delta}_\rho(t) = te^{-3(1+\rho)}.$$

Therefore (6.6) reads

$$\sup_{t\in[\frac{1}{8},\frac{1}{4}]} \left\{ e^{-3(1+\rho)} \int_{\frac{1}{8}}^{\frac{1}{4}} k(t,s)s\,ds \right\} \geq e^{-3(1+\rho)} \int_{\frac{1}{8}}^{\frac{1}{4}} \frac{1}{16} s^2\,ds = \frac{7e^{-3(1+\rho)}}{24576} > 0,$$

which implies that (6.6) is satisfied for every $\rho \in (0, +\infty)$.

Thus we can apply Theorem 6.3 obtaining uncountably many pairs of solutions and parameters (u_ρ, λ_ρ) for the BVP (6.15). □

Acknowledgments The authors would like to thank the referee for the careful reading of the manuscript and the constructive comments. The authors were partially supported by the Gruppo Nazionale per l'Analisi Matematica, la Probabilità e le loro Applicazioni (GNAMPA) of the Istituto Nazionale di Alta Matematica (INdAM). G. Infante is a member of the UMI Group TAA "Approximation Theory and Applications." This paper was partially written during the visit of A. Calamai to the Dipartimento di Matematica e Informatica of the Università della Calabria. A. Calamai is grateful to the people of the aforementioned Dipartimento for their kind and warm hospitality.

References

1. A. Cabada, G. Infante, F.A.F. Tojo, Nonzero solutions of perturbed Hammerstein integral equations with deviated arguments and applications. Topol. Methods Nonlinear Anal. **47**, 265–287 (2016)
2. A. Calamai, G. Infante, Nontrivial solutions of boundary value problems for second order functional differential equations. Ann. Mat. Pura Appl. **195**, 741–756 (2016)
3. A. Calamai, G. Infante, An affine Birkhoff–Kellogg type result in cones with applications to functional differential equations. Math. Methods Appl. Sci. **46**(11), 11897–11905 (2023). https://doi.org/10.1002/mma.8665
4. A. Calamai, G. Infante, On fourth order retarded equations with functional boundary condition: a unified approach. Discrete Contin. Dyn. Syst. Ser. S **17**, 2009–2020 (2024). https://doi.org/10.3934/dcdss.2023006
5. F. Cianciaruso, G. Infante, P. Pietramala, Solutions of perturbed Hammerstein integral equations with applications. Nonlinear Anal. Real World Appl. **33**, 317–347 (2017)

6. S. Djebali, K. Mebarki, Fixed point index on translates of cones and applications. Nonlinear Stud. **21**, 579–589 (2014)
7. H. Fan, R. Ma, Loss of positivity in a nonlinear second order ordinary differential equations. Nonlinear Anal. **71**, 437–444 (2009)
8. R. Figueroa, R.L. Pouso, Minimal and maximal solutions to second-order boundary value problems with state-dependent deviating arguments. Bull. Lond. Math. Soc. **43**, 164–174 (2011)
9. D. Franco, G. Infante, J. Perán, A new criterion for the existence of multiple solutions in cones. Proc. Roy. Soc. Edinburgh Sect. A **142**, 1043–1050 (2012)
10. G. Infante, Eigenvalues of some non-local boundary-value problems. Proc. Edinb. Math. Soc. **46**, 75–86 (2003)
11. G. Infante, Positive solutions of differential equations with nonlinear boundary conditions. Discrete Contin. Dyn. Syst. **Suppl. Vol. 2003**, 432–438 (2003)
12. G. Infante, Nonlocal boundary value problems with two nonlinear boundary conditions. Commun. Appl. Anal. **12**, 279–288 (2008)
13. G. Infante, J.R.L. Webb, Three point boundary value problems with solutions that change sign. J. Integral Equations Appl. **15**, 37–57 (2003)
14. G. Infante, J.R.L. Webb, Nonlinear nonlocal boundary value problems and perturbed Hammerstein integral equations. Proc. Edinb. Math. Soc. **49**, 637–656 (2006)
15. G. Infante, J.R.L. Webb, Loss of positivity in a nonlinear scalar heat equation. NoDEA Nonlinear Differential Equations Appl. **13**, 249–261 (2006)
16. G. Infante, J.R.L. Webb, Nonlinear non-local boundary-value problems and perturbed Hammerstein integral equations. Proc. Edinb. Math. Soc. **49**, 637–656 (2006)
17. G. Kalna, S. McKee, The thermostat problem. TEMA Tend. Mat. Apl. Comput. **3**, 15–29 (2002)
18. G. Kalna, S. McKee, The thermostat problem with a nonlocal nonlinear boundary condition. IMA J. Appl. Math. **69**, 437–462 (2004)
19. I. Karatsompanis, P.K. Palamides, Polynomial approximation to a non-local boundary value problem. Comput. Math. Appl. **60**, 3058–3071 (2010)
20. M.A. Krasnosel'skiĭ, L.A. Ladyženskiĭ, The structure of the spectrum of positive nonhomogeneous operators. Trudy Moskov. Mat. Obšč **3**, 321–346 (1954)
21. J.J. Nieto, J. Pimentel, Positive solutions of a fractional thermostat model. Bound. Value Probl. **2013**(5), 11 pp. (2013)
22. P. Palamides, G. Infante, P. Pietramala, Nontrivial solutions of a nonlinear heat flow problem via Sperner's Lemma. Appl. Math. Lett. **22**, 1444–1450 (2009)
23. C. Shen, H. Zhou and L. Yang, Existence of positive solutions of a nonlinear differential equation for a thermostat model. Math. Methods Appl. Sci. **41**, 6145–6154 (2018)
24. J.R.L. Webb, Multiple positive solutions of some nonlinear heat flow problems. Discrete Contin. Dyn. Syst. (Suppl.) **2005**, 895–903 (2005)
25. J.R.L. Webb, Optimal constants in a nonlocal boundary value problem. Nonlinear Anal. **63**, 672–685 (2005)
26. J.R.L. Webb, Existence of positive solutions for a thermostat model. Nonlinear Anal. Real World Appl. **13**, 923–938 (2012)

Chapter 7
Branches of Forced Oscillations for a Class of Implicit Equations Involving the Φ-Laplacian

Alessandro Calamai, Maria Patrizia Pera, and Marco Spadini

MSC 2010 34C25, 34A09, 34B08, 47H11

7.1 Introduction

In this chapter we study, by topological methods, the forced oscillations of some nonlinear, implicit ordinary differential equations (ODEs) governed by a generalized Φ-Laplacian-type differential operator.

Implicit ODEs arise naturally in applications and have been studied by many authors. For this kind of equations, usually the existence of homoclinic/heteroclinic orbits has been investigated with different methods, for example, with a dynamical system approach (see, e.g., the recent [3] and the survey [20]). As far as we know, instead, not many papers have been devoted to the study of periodic solutions.

A typical example, in the scalar case, of equations for which it is not possible, or convenient, to write them explicitly is provided by those involving the classical p-Laplacian operator $\Phi(x) := x|x|^{p-2}$, with $p > 1$, or, more generally, the so-called Φ-*Laplacian*: meaning a strictly increasing homeomorphism $\Phi : \mathbb{R} \to \mathbb{R}$. That is, scalar equations of the form

The authors are members of the Gruppo Nazionale per l'Analisi Matematica, la Probabilità e le loro Applicazioni (GNAMPA) of the Istituto Nazionale di Alta Matematica (INdAM)

A. Calamai (✉)
Dipartimento di Ingegneria Civile, Edile e Architettura, Università Politecnica delle Marche Via Brecce Bianche, Ancona, Italy
e-mail: a.calamai@univpm.it

M. P. Pera · M. Spadini
Dipartimento di Matematica e Informatica "Ulisse Dini", Università degli Studi di Firenze, Florence, Italy
e-mail: mpatrizia.pera@unifi.it; marco.spadini@unifi.it

P. Amster, P. Benevieri (eds.), *Topological Methods for Delay and Ordinary Differential Equations*, Advances in Mechanics and Mathematics 51,
https://doi.org/10.1007/978-3-031-61337-1_7

$$(\Phi(x'))' = f(t, x, x').$$

These equations are well-studied since they arise in some applicative models: for example, in non-Newtonian fluid theory, nonlinear elasticity, diffusion of flows in porous media, theory of capillary surfaces, and, more recently, the modeling of glaciology (see, e.g., [4, 9, 32]). From a mathematical viewpoint, different kinds of boundary conditions can be associated to these equations; for example, the existence of heteroclinic solutions can be established using the method of upper/lower solutions (see, e.g., [7, 11, 13, 16, 18]). The study of periodic solutions has also been pursued in $\mathbb{R}^n$ via degree theory: We cite, for instance [5, 6, 10, 19, 21, 22, 33].

Here we study periodic solutions of some implicit ODEs in $\mathbb{R}^n$ with generalized Φ-Laplacian-type terms. We follow a topological approach based on the Brouwer degree. We consider two different problems that have to be handled separately. Namely,

$$[\phi_\lambda\big(x(t), x'(t)\big)]' = \lambda f\big(t, x(t), x'(t)\big), \quad \lambda \geq 0, \tag{7.1}$$

and

$$[\phi_\lambda\big(x(t), x'(t)\big)]' = g\big(x(t), x'(t)\big) + \lambda f\big(t, x(t), x'(t)\big), \quad \lambda \geq 0, \tag{7.2}$$

where λ is a real parameter and, for a given open set $U \subseteq \mathbb{R}^n$, the maps $f: \mathbb{R} \times U \times \mathbb{R}^n \to \mathbb{R}^n$ and $g: U \times \mathbb{R}^n \to \mathbb{R}^n$ are continuous. We assume that f is T-periodic in the first variable, $T > 0$ being fixed, and we look for periodic solutions of (7.1), resp. (7.2), of the same period T.

Concerning the left-hand side in (7.1) and (7.2), we assume that for each $\lambda \in [0, \infty)$ and $x \in U$, the map

$$\phi_\lambda(x, \cdot) : \mathbb{R}^n \to \mathbb{R}^n$$

is one to one and onto. Therefore, with the introduction of an auxiliary variable $y(t) = \phi_\lambda\big(x(t), x'(t)\big)$, problems (7.1) and (7.2) can be equivalently written as explicit systems in $\mathbb{R}^{2n}$. As a consequence, the properties of the solution sets of (7.1) and (7.2) can be investigated by means of earlier results, obtained by the authors also in collaboration with Massimo Furi, about periodically perturbed ODEs on differentiable manifolds (see [24, 26, 35]). We observe that the mentioned results rely upon the topological notion of degree (also called *Euler characteristic*) of a tangent vector field, see the classical books [28–30]. In spite of this fact, no advanced tools from differential topologies are needed in the present paper. In fact, the technicalities are hidden in the proofs, and our main results require only the well-known Brouwer degree in Euclidean spaces.

In our main results, Theorems 7.4 and 7.5, we investigate qualitative properties of the set of T-periodic solutions of (7.1) and (7.2). Roughly speaking, we obtain the existence of an unbounded connected set—a "branch"—made up of "nontrivial" T-periodic solution pairs (λ, x) of (7.1) and (7.2) that emanates from the set of the "trivial" ones—see Sect. 7.2 for more precise definitions. Thus, we may describe our main theorems as *global bifurcation results*. To prove the existence of such a branch, we need only suitable assumptions on the Brouwer degree of some maps in $\mathbb{R}^n$, related with the terms f and g in the right-hand side of (7.1) and (7.2). Observe that the Brouwer degree can be computed explicitly, and so, in principle, it is possible, and often not so difficult, to verify the validity of those assumptions.

Our results are somehow related to similar ones, concerning the set of T-periodic solutions of periodically perturbed differential–algebraic equations, recently obtained by the authors [8, 12, 15, 36]. In fact, under suitable conditions, also in that case it is possible to relate such equations to ODEs on manifolds and to obtain the desired properties only in terms of the Brouwer degree in $\mathbb{R}^n$.

Another context in which the abovementioned results on periodically perturbed ODEs on manifolds have been successfully exploited is that of delay differential equations (see, e.g., [14]). We think that the results obtained here could be generalized to equations in which a dependence on delayed terms is introduced (see also [1]). However we do not pursue here this line: See Sect. 7.5 for a brief discussion.

Observe finally that our results are not directly deducible from the standard implicit function theorem. And this for different reasons. Firstly, the implicit function theorem provides information on local properties, while the results we obtain here are of global nature. Next, the application of the implicit function theorem requires more regularity than that we assume here (i.e., the involved maps need to be of class C^1 and not merely continuous). In fact, apart from a technical assumption on the partial differentiability with respect to the parameter λ, we only require continuity. We believe that our global bifurcation results could be even generalized to a Carathéodory-type setting. This will be a further line of future investigation.

7.2 Setting of the Problem

Assume throughout the chapter that $\phi\colon [0,\infty) \times U \times \mathbb{R}^n \to \mathbb{R}^n$, U being open in $\mathbb{R}^n$, is a continuous map with the property that for each $\lambda \in [0,\infty)$ and $u \in U$, the map

$$\phi(\lambda, u, \cdot) : \mathbb{R}^n \to \mathbb{R}^n$$

is one to one and onto; in other words, $\phi(\lambda, u, \cdot)$ is invertible with respect to the third variable in the following sense: For each $w \in \mathbb{R}^n$ there exists a unique $v \in \mathbb{R}^n$, depending on (λ, u), such that

$$w = \phi(\lambda, u, v).$$

We will denote this "partial inverse" as $\psi(\lambda, u, \cdot)$, so that $v = \psi(\lambda, u, w)$. We will also assume that, for each $u \in U$ and $w \in \mathbb{R}^n$, the partial derivative

$$\lambda \mapsto \partial_1 \psi(\lambda, u, w) \quad \text{exists and is continuous.} \tag{7.3}$$

Here and in what follows, the symbol "∂_i" denotes the partial differentiation with respect to the i-th variable.

We explicitly observe that, throughout the chapter, very weak regularity assumptions are needed on ϕ, which is merely assumed to be continuous and such that (7.3) holds. In fact, the map ϕ will be assumed to be a perturbation of a homeomorphism; however, in the more regular case, the following remark is in order.

Remark 7.1 If the map ϕ is of class C^1, by applying the implicit function theorem, one gets that the above condition (7.3) is satisfied if, for any $(\lambda, u, v) \in [0, \infty) \times U \times \mathbb{R}^n$, the partial derivative

$$\partial_3 \phi(\lambda, u, v)$$

is invertible.

It is sometimes convenient to write $\phi_\lambda(u, v)$ instead of $\phi(\lambda, u, v)$. Finally, we suppose that $\phi(0, \cdot, \cdot)$ depends only on the third variable, that is, by abuse of notation, $\phi(0, u, v) =: \phi_0(v)$ and thus we may write $\psi(0, u, w) = \phi_0^{-1}(w)$. The latter assumption means that we regard ϕ_λ as a perturbation of some nonlinearity depending only on the third variable. With this in mind, we view $\phi(\lambda, u, v)$ as a sum $\phi_0(v) + \lambda\delta(\lambda, u, v)$, where ϕ_0 is a homeomorphism and $\delta(\lambda, u, v)$ is a perturbation such that the above assumptions are satisfied.

A simple example of a map ϕ as above, in the case $n = 1$, is the following:

$$\phi(\lambda, u, v) = (v + \lambda)|v + \lambda|^{p-2}, \quad p = 2, 3, \ldots,$$

which is a perturbation of the classical p-Laplacian, namely $\phi_0(v) = v|v|^{p-2}$. In this case one can explicitly compute, for $p = 2$, $\psi(\lambda, u, w) = w - \lambda$, while for $p > 2$,

$$\psi(\lambda, u, w) = \operatorname{sign}(w)|w|^{1/(p-1)} - \lambda,$$

and in any case we get $\partial_1 \psi(\lambda, u, w) = -1$ so that the above assumptions are satisfied.

As another example in the case $n = 1$, consider

$$\phi(\lambda, u, v) = \Phi(v) + \lambda(v - u),$$

where $\Phi : \mathbb{R} \to \mathbb{R}$, a Φ-Laplacian-type operator, is a diffeomorphism such that $\Phi'(v) > 0$ for any $v \in \mathbb{R}$. Even in this case one can check that the above assumptions are satisfied.

We wish to investigate the two following problems that, although somewhat similar, lead to substantially different results:

$$[\phi_\lambda\big(x(t),x'(t)\big)]' = \lambda f\big(t,x(t),x'(t)\big), \quad \lambda \geq 0, \tag{7.4}$$

and

$$[\phi_\lambda\big(x(t),x'(t)\big)]' = g\big(x(t),x'(t)\big) + \lambda f\big(t,x(t),x'(t)\big), \quad \lambda \geq 0, \tag{7.5}$$

where $f\colon \mathbb{R}\times U\times\mathbb{R}^n \to \mathbb{R}^n$ and $g\colon U\times\mathbb{R}^n\to\mathbb{R}^n$ are continuous maps and f is assumed T-periodic in the first variable for a $T>0$ given.

By a T-periodic solution of (7.4), corresponding to a given $\lambda \geq 0$, we mean a C^1 and T-periodic function $x\colon \mathbb{R}\to\mathbb{R}^n$ such that there exists another C^1 and T-periodic function $y\colon\mathbb{R}\to\mathbb{R}^n$ such that the system

$$\begin{cases} x'(t) = \psi\big(\lambda, x(t), y(t)\big), \\ y'(t) = \lambda f\Big(t, x(t), \psi\big(\lambda, x(t), y(t)\big)\Big), \end{cases} \tag{7.6}$$

is satisfied identically. The T-periodic solutions of (7.5), corresponding to $\lambda \geq 0$, are defined analogously as the T-periodic solutions of the system

$$\begin{cases} x'(t) = \psi\big(\lambda, x(t), y(t)\big), \\ y'(t) = g\Big(x(t), \psi\big(\lambda, x(t), y(t)\big)\Big) + \lambda f\Big(t, x(t), \psi\big(\lambda, x(t), y(t)\big)\Big). \end{cases} \tag{7.7}$$

Observe that from the first equation in (7.6) we get $\phi_\lambda\big(x(t),x'(t)\big) = y(t)$; hence the map $t \mapsto \phi_\lambda\big(x(t),x'(t)\big)$ is C^1. Finally, notice that when the partial derivatives of ϕ_λ exist continuous along the curve $t\mapsto \big(x(t),x'(t)\big)\in\mathbb{R}^{2n}$ with $\partial_2\phi_\lambda\big(x(t),x'(t)\big)$ invertible, one has that x is actually C^2 since, in this case, we have

$$x''(t) = \partial_2\phi_\lambda\big(x(t),x'(t)\big)^{-1}\Big[\lambda f\big(t,x(t),x'(t)\big) - \partial_1\phi_\lambda\big(x(t),x'(t)\big)x'(t)\Big].$$

We denote by $C^r_T(U)$ the set of the U-valued, T-periodic, C^r functions with topology induced by the C^r norm. This is a subset of the Banach space $C^r_T(\mathbb{R}^n)$. For simplicity, when $r=0$, we write $C_T(U)$ instead of $C^0_T(U)$.

A pair $(\lambda,x)\in[0,\infty)\times C^1_T(U)$ with x taking values in U, such that x is a T-periodic solution of (7.4) (resp., (7.5)), is said to be a *T-forced pair* for (7.4) (resp., (7.5)). A T-forced pair (λ,x) is called *trivial* if x is constant and $\lambda=0$. In this chapter we investigate the set of nontrivial T-forced pairs of (7.4) and (7.5).

7.3 Preliminaries

In order to pursue our investigation on the set of T-forced pairs of (7.4) and (7.5) we need slight generalizations of previous results, obtained in [24, 26, 35]. These results deal with periodic perturbations of autonomous ODEs on manifolds and require, in the statements, the notion of degree (also called *Euler characteristic*) of a tangent vector field.

In this preliminary section we first recall this notion of degree and its relationship with the Brouwer degree. As already pointed out, in this chapter the degree is only a tool, whose use is hidden in the proofs. In spite of this, we decided to add the definition and some properties of the degree on manifolds in order to make the paper self-contained. The reader who is already familiar with it can skip Sect. 7.3.1.

7.3.1 The Degree of a Tangent Vector Field

Let M be a smooth boundaryless submanifold of $\mathbb{R}^k$ and $w\colon M \to \mathbb{R}^k$ a continuous tangent vector field on M: meaning that, for all $p \in M$, we have $w(p) \in T_pM$, where $T_pM \subseteq \mathbb{R}^k$ is the tangent space of M at p. Particularly relevant in this chapter is the special case when M is an open subset of $\mathbb{R}^k$. When this happens, one has $T_pM = \mathbb{R}^k$ for all $p \in \mathbb{R}^k$.

Let $U \subseteq M$ be open. The pair (w, U) is said to be *admissible* if $w^{-1}(0) \cap U$ is compact; we also say that w is admissible for the degree in U. It is known (see, e.g., [28–30]) that one can associate to any admissible pair (w, U) an integer, $\deg(w, U)$, called the *degree* of w in U. The degree, roughly speaking, counts algebraically the number of zeros of w in U.

When w is (Fréchet) differentiable at $p \in M$ and $w(p) = 0$, then its derivative at p, $w'(p)\colon T_pM \to \mathbb{R}^k$, actually maps T_pM into itself, and, therefore, its determinant $\det w'(p)$ is well-defined. If, moreover, p is a nondegenerate zero (i.e., $w'(p)\colon T_pM \to \mathbb{R}^k$ is injective), then p is an isolated zero and $\det w'(p) \neq 0$.

In the regular case, that is, when w is admissible for the degree in U and the zeros of w are all nondegenerate, then the set $w^{-1}(0) \cap U$ is finite and the following equality holds:

$$\deg(w, U) = \sum_{p \in w^{-1}(0) \cap U} \operatorname{sign} \det w'(p). \tag{7.8}$$

Moreover, we stress that when $M = \mathbb{R}^k$, $\deg(w, U)$ coincides with the Brouwer degree, $\deg_B(w, V, 0)$, of the triple $(w, V, 0)$: where $V \subseteq U$ is any bounded open set, whose closure is contained in U, containing the compact set $w^{-1}(0) \cap U$, i.e.,

$$\deg_B(w, V, 0) = \deg(w, U). \tag{7.9}$$

All the standard properties of the classical Brouwer degree on open and bounded subsets of Euclidean spaces, such as homotopy invariance, excision, additivity, existence, are still valid in the more general context of differentiable manifolds.

Indeed, it can be shown (see [25]) that the degree of a tangent vector field is uniquely determined by some of these properties used as axioms.

The classical Brouwer degree can be extended to the case when the open set U is possibly unbounded, see, e.g., [17, Sec. 1.2.4] or, in the context of differentiable manifolds, [31, Sec. 1.3]. It is not difficult to show, using the excision property, that (7.9) is valid also in this extended context. Indeed, one could use (7.9) as basis for the extended notion of Brouwer degree.

In this chapter, we will always omit the "target" point in the notation for the degree since we are only interested in the case when this is 0. By abuse of notation justified by (7.9), we will write deg to denote both *the extended Brouwer degree* and *the degree of a tangent vector field.*

7.3.2 *On the Set of T-Periodic Solutions of Perturbed ODEs on Manifolds*

In this section we will be concerned with differential equations on manifolds. We need to introduce some further notation. Let $M \subseteq \mathbb{R}^k$ be a boundaryless differentiable manifold. By $C_T(M)$ we denote the set of the M-valued, T-periodic, continuous functions with the topology induced by the Banach space $C_T(\mathbb{R}^k)$.

Consider the following differential equations on $M \subseteq \mathbb{R}^k$ depending on a parameter $\lambda \in [0, \infty)$:

$$\dot{\xi}(t) = \mathfrak{g}\big(\xi(t)\big) + \lambda \mathfrak{f}\big(t, \xi(t), \lambda\big), \tag{7.10}$$

and

$$\dot{\xi}(t) = \lambda \mathfrak{f}\big(t, \xi(t), \lambda\big), \tag{7.11}$$

where $\mathfrak{f}\colon \mathbb{R} \times M \times [0, \infty) \to \mathbb{R}^k$ and $\mathfrak{g}\colon M \to \mathbb{R}^k$ are continuous vector fields tangent to M, $\mathfrak{f}$ being T-periodic in the first variable.

A pair $(\lambda, \xi) \in [0, \infty) \times C_T(M)$, such that (7.10) (resp., (7.11)) holds identically, is a *T-pair* for (7.10) (resp., (7.11)). A T-pair (λ, ξ) is called *trivial* if ξ is constant and $\lambda = 0$. Below, we concentrate on the set of nontrivial T-pairs of (7.10) and (7.11).

Given any point $p \in M$, it is convenient to introduce the notation $\overline{p}$ to denote the function constantly equal to p. Accordingly, a T-pair for (7.10) is trivial if and only if it is of the form $(0, \overline{p})$ for some $p \in \mathfrak{g}^{-1}(0)$. Similarly, for all $p \in M$, all pairs $(0, \overline{p})$ are trivial T-pairs for (7.11).

The qualitative properties of the set of T-pairs of (7.10) and (7.11) can be deduced by suitable assumptions on the degree of some tangent vector fields, see [26].

In fact, by inspection of Theorem 3.3 of [26] one immediately sees that the following result holds:

Theorem 7.1 *Let $\mathfrak{f}$ and $\mathfrak{g}$ be as in Eq.* (7.10) *and let Ω be an open subset of $[0,\infty)\times C_T(M)$. Put $\Omega_M = \{p \in M : (0,\overline{p}) \in \Omega\}$. Assume that* $\deg(\mathfrak{g},\Omega_M)$ *is well-defined and nonzero. Then there exists a connected set Γ of nontrivial T-pairs in Ω whose closure in $[0,\infty)\times C_T(M)$ intersects the set $\{(0,\overline{p}) \in [0,\infty)\times C_T(M) : p \in \mathfrak{g}^{-1}(0)\cap\Omega_M\}$ and is not contained in any compact subset of Ω. In particular, if M is closed in $\mathbb{R}^k$ and $\Omega = [0,\infty)\times C_T(M)$, then Γ is unbounded.*

Similarly, by inspection of the proof of Theorem 2.3 of [24] one obtains the following:

Theorem 7.2 *Let $\mathfrak{f}$ be as in* (7.11), *and define $\mathfrak{v}\colon M \to \mathbb{R}^k$ be the autonomous vector field given by*

$$\mathfrak{v}(p) := \frac{1}{T}\int_0^T \mathfrak{f}(t,p,0)\,\mathrm{d}t.$$

Let Ω be an open subset of $[0,\infty)\times C_T(M)$, and put $\Omega_M = \{p \in M : (0,\overline{p}) \in \Omega\}$. Assume that $\deg(\mathfrak{v},\Omega_M)$ *is well-defined and nonzero. Then there exists a connected set Γ of nontrivial T-pairs in Ω whose closure in $[0,\infty)\times C_T(M)$ meets $\{(0,\overline{p}) \in [0,\infty)\times C_T(M) : p \in \mathfrak{v}^{-1}(0)\cap\Omega_M\}$ and is not contained in any compact subset of Ω. In particular, if M is closed in $\mathbb{R}^k$ and $\Omega = [0,\infty)\times C_T(M)$, then Γ is unbounded.*

Observe that Theorem 7.2 concerns the case not covered in general by Theorem 7.1 of perturbations of the zero vector field. In fact, unless the manifold M is compact, Theorem 7.1 above is not applicable, $\mathfrak{g}$ not being admissible.

Finally, consider the following system of coupled equations, depending on the parameter $\lambda \geq 0$, on the manifold $M \times N$, where $M \subseteq \mathbb{R}^k$ and $N \subseteq \mathbb{R}^s$ are differentiable manifolds

$$\begin{cases} \dot{\xi}(t) = \lambda\mathfrak{f}\big(t,\xi(t),\eta(t),\lambda\big), \\ \dot{\eta}(t) = \mathfrak{g}\big(\xi(t),\eta(t)\big) + \lambda\mathfrak{h}\big(t,\xi(t),\eta(t),\lambda\big), \end{cases} \tag{7.12}$$

where $\mathfrak{f}\colon \mathbb{R}\times M\times N\times[0,\infty) \to \mathbb{R}^k$, $\mathfrak{h}\colon \mathbb{R}\times M\times N\times[0,\infty) \to \mathbb{R}^s$, and $\mathfrak{g}\colon M\times N \to \mathbb{R}^s$ are continuous vector fields tangent to $M\times N$, $\mathfrak{f}$, and $\mathfrak{h}$ being T-periodic in the first variable.

A triple $(\lambda,x,y) \in [0,\infty)\times C_T(M\times N)$, such that (7.12) holds identically, is a *T-triple* for (7.12). A T-triple (λ,x,y) is called *trivial* if (x,y) is constant and $\lambda = 0$.

As above, given $(p,q) \in M\times N$, by $\overline{p}$ and $\overline{q}$, we denote the functions constantly equal to p and q, respectively. Thus, a T-triple is trivial if and only if it is of the form $(0,\overline{p},\overline{q})$ with $(p,q) \in \mathfrak{g}^{-1}(0)$.

Let $\nu : M \times N \to \mathbb{R}^{k+s}$ be the vector field, tangent to the manifold $M \times N \subseteq \mathbb{R}^{k+s}$, given by

$$\nu(p, q) = \left(\frac{1}{T} \int_0^T \mathfrak{f}(t, p, q, 0)\, \mathrm{d}t \ , \ \mathfrak{g}(p, q) \right).$$

As for the previous two theorems, an inspection of the argument of Theorem 4 of [35] shows that the following statement holds:

Theorem 7.3 *Let $\mathfrak{f}$, $\mathfrak{g}$, and $\mathfrak{h}$ be as in Eq. (7.12), and let Ω be an open subset of $[0, \infty) \times C_T(M \times N)$. Put $\Omega_{M\times N} = \{(p, q) \in M \times N : (0, \overline{p}, \overline{q}) \in \Omega\}$. Assume that $\deg(\nu, \Omega_{M\times N})$ is well-defined and nonzero. Then there exists a connected set Γ of nontrivial T-triples in Ω of (7.12) whose closure in $[0, \infty) \times C_T(M \times N)$ intersects $\{(0, \overline{p}, \overline{q}) \in [0, \infty) \times C_T(M \times N) : (p, q) \in \nu^{-1}(0) \cap \Omega_{M\times N}\}$ and is not contained in any compact subset of Ω. In particular, if $M \times N$ is closed in $\mathbb{R}^{k+s}$ and $\Omega = [0, \infty) \times C_T(M \times N)$, then Γ is unbounded.*

7.4 Main Results

In this section we investigate the set of T-pairs for Eqs. (7.4) and (7.5). A crucial step is the following simple observation that essentially is a version of Hadamard's Lemma, see, e.g., [2]:

Remark 7.2 Let ϕ be as in Eqs. (7.4) and (7.5). Let also $u \in U$, $v, w \in \mathbb{R}^n$, and $\lambda \in [0, \infty)$ be such that $w = \phi_\lambda(u, v)$; then there exists a continuous function h such that $v = \phi_0^{-1}(w) + \lambda h(\lambda, u, w)$. To see that this is the case, let us write $v = \psi(\lambda, u, w)$, and then, in order to isolate λ, we observe that

$$\begin{aligned} v - \phi_0^{-1}(w) = v - \psi(0, u, w) &= \int_0^1 \frac{\mathrm{d}}{\mathrm{d}s} \psi(s\lambda, u, w)\, \mathrm{d}s \\ &= \int_0^1 \lambda \partial_1 \psi(s\lambda, u, w)\, \mathrm{d}s = \lambda h(\lambda, u, w), \end{aligned}$$

where $h(u, w, \lambda) = \int_0^1 \partial_1 \psi(s\lambda, u, w)\, \mathrm{d}s$.

We consider separately the cases of Eqs. (7.4) and (7.5).

7.4.1 *On the Set of T-Forced Pairs for Equation (7.4)*

Introducing the new function $y(t) = \phi_\lambda(x(t), x'(t))$ and using Remark 7.2, we are led to the following system:

$$\begin{cases} x'(t) = \phi_0^{-1}\big(y(t)\big) + \lambda h(\lambda, x(t), y(t)), \\ y'(t) = \lambda f\Big(t, x(t), \phi_0^{-1}\big(y(t)\big) + \lambda h(\lambda, x(t), y(t))\Big), \end{cases} \tag{7.13}$$

which is of the form (7.12). We relate the set $X \subseteq [0, \infty) \times C_T^1(U)$ of T-forced pairs of (7.4) with the set $Y \subseteq [0, \infty) \times C_T(U \times \mathbb{R}^n)$ of T-triples of (7.13).

Define the map $\mathfrak{H}\colon Y \to X$ by $\mathfrak{H}\colon (\lambda, x, y) \mapsto (\lambda, x)$, and observe that $\mathfrak{H}$ is continuous.

Lemma 7.1 *The map $\mathfrak{H}$ is a homeomorphism that respects the notion of triviality, in the sense that it makes trivial T-triples correspond to trivial T-forced pairs and vice versa.*

Proof First notice that if (λ, x, y) is a T-triple of (7.13), then x is a T-periodic solution of (7.4) corresponding to λ. In particular, when (λ, x, y) is trivial, $\lambda = 0$ and x is constant: Hence (λ, x) is a trivial T-forced pair.

Conversely, take any $(\lambda, x) \in X$, and let $y(t) = \phi_\lambda\big(x(t), x'(t)\big)$. We immediately see that y is T-periodic and that (x, y) is a T-periodic solution of (7.13) corresponding to λ. In other words

$$\mathfrak{H}^{-1}(\lambda, x) = \big(\lambda, x, \phi_\lambda\big(x, x'\big)\big) = (\lambda, x, y),$$

and this map is obviously continuous. Observe, in particular, that when x is constant so is y, hence the image under $\mathfrak{H}^{-1}$ of any trivial T-pair is a trivial T-forced pair. □

Consider the following vector field, defined on $U \times \mathbb{R}^n$:

$$\nu(p, q) := \left(\frac{1}{T}\int_0^T f\big(t, p, \phi_0^{-1}(q)\big)\, \mathrm{d}t \;,\; \phi_0^{-1}(q)\right).$$

Taking into account Theorem 7.3 and Lemma 7.1, one sees that the degree of ν plays a crucial role in our investigation. Furthermore, since ϕ_0^{-1} is a homeomorphism, the computation of the degree of ν can be reduced to the degree of the "average wind":

$$w(p) := \frac{1}{T}\int_0^T f(t, p, 0)\, \mathrm{d}t, \tag{7.14}$$

as shown by the following technical lemma:

Lemma 7.2 *Let $W \subseteq U$ be open. The vector field ν is admissible for the degree in $W \times \mathbb{R}^n$ if and only if so is w in W, and*

$$\deg\big(\nu, W \times \mathbb{R}^n\big) = \pm \deg\big(w, W\big). \tag{7.15}$$

Proof Since ϕ_0^{-1} is a homeomorphism, let $q_0 \in \mathbb{R}^n$ be the (unique) point such that $\phi_0^{-1}(q_0) = 0$. We have

$$\nu^{-1}(0) = \big\{(p, q_0) : p \in w^{-1}(0)\big\}, \tag{7.16}$$

so that $w^{-1}(0) \cap W$ is compact if and only if so is $\nu^{-1}(0) \cap (W \times \mathbb{R}^n)$. This implies that ν is admissible in $W \times \mathbb{R}^n$ if and only if so is w in W.

We now prove formula (7.15). Let W_1 and W_2 be open and bounded subsets of W and $\mathbb{R}^n$ respectively such that $\nu^{-1}(0) \subseteq W_1 \times W_2$. By the excision property of the degree, we find

$$\deg\big(\nu, W_1 \times \mathbb{R}^n\big) = \deg\big(\nu, W_1 \times W_2\big). \tag{7.17}$$

Denote by ν_1 the first component of ν. By known transversality theorems, we can approximate ν_1 with a smooth map $\hat{\nu}_1$ with only isolated zeros and also approximate ϕ_0 with a smooth diffeomorphism $\hat{\phi}_0$. Let

$$\hat{\nu}(p, q) = \big(\hat{\nu}_1(p, q)\,,\ \hat{\phi}_0^{-1}(q)\big).$$

We can take the approximations so close that the homotopy

$$\begin{aligned} H(s, p, q) &= s\hat{\nu}(p, q) + (1 - s)\nu(p, q) \\ &= \Big(s\hat{\nu}_1(p, q) + (1 - s)\nu_1(p, q)\,,\ s\hat{\phi}_0^{-1}(q) + (1 - s)\phi_0^{-1}(q)\Big), \end{aligned}$$

has no zeros on the boundary of the bounded open set $W_1 \times W_2$ for $s \in [0, 1]$. Thus H is admissible in $W_1 \times W_2$. The homotopy property of the degree yields

$$\deg\big(\nu, W_1 \times W_2\big) = \deg\big(\hat{\nu}, W_1 \times W_2\big). \tag{7.18}$$

Similarly, defining $\hat{w}(p) = \hat{\nu}_1(p, q_0)$, we have

$$\deg(w, W_1) = \deg(\hat{w}, W_1). \tag{7.19}$$

Given an isolated zero (p_0, q_0) of $\hat{\nu}$, we have

$$\det \hat{\nu}'(p_0, q_0) = \det \begin{pmatrix} \partial_1 \hat{\nu}_1(p_0, q_0) & 0 \\ \partial_2 \hat{\nu}_1(p_0, q_0) & (\hat{\phi}_0^{-1})'(q_0) \end{pmatrix} = \det \hat{w}'(p_0) \det(\phi_0^{-1})'(q_0).$$

As in (7.16) we have

$$\hat{\nu}^{-1}(0) = \big\{(p, q_0) : p \in \hat{w}^{-1}(0)\big\}.$$

Also, $\hat{\phi}_0^{-1}$ being a diffeomorphism, $\operatorname{sign}\det(\hat{\phi}_0^{-1})'(q_0) = \pm 1$. Thus,

$$\begin{aligned}\deg(\hat{\nu}, W_1 \times W_2) &= \sum_{(p_0,q_0)\in\hat{\nu}^{-1}(0)\cap W_1\times W_2} \operatorname{sign}\det \hat{\nu}'(p_0, q_0) \\ &= \operatorname{sign}\det(\hat{\phi}_0^{-1})'(q_0) \sum_{p_0\in\hat{w}^{-1}(0)\cap W_1} \operatorname{sign}\det \hat{w}'(p_0) \\ &= \pm\deg(\hat{w}, W_1).\end{aligned}$$

The assertion now follows from Eqs. (7.19), (7.18), and (7.17). □

Given an open subset Ω of $[0,\infty) \times C_T(U)$, the set $\Omega^* = \Omega \times C_T(\mathbb{R}^n)$ is open in $[0,\infty) \times C_T(U \times \mathbb{R}^n)$. As in Sect. 7.3 we define

$$\Omega_U = \left\{p \in U : (0, \overline{p}) \in \Omega\right\}$$

that is clearly open in U. Similarly, we consider the open set $\Omega^*_{U\times\mathbb{R}^n} \subseteq U \times \mathbb{R}^n$ given by

$$\Omega^*_{U\times\mathbb{R}^n} = \left\{(p, q) \in U \times \mathbb{R}^n : (0, \overline{p}, \overline{q}) \in \Omega^*\right\} = \Omega_U \times \mathbb{R}^n.$$

By Lemma 7.2 we have that when $\deg(w, \Omega_U)$ is well-defined and nonzero, then so is $\deg(\nu, \Omega^*_{U\times\mathbb{R}^n})$. Thus, by Theorem 7.3 applied to (7.13), we find that there exists a connected set, say Θ, of nontrivial T-triples in Ω of (7.13) whose closure in $[0,\infty) \times C_T(U \times \mathbb{R}^n)$ meets $\nu^{-1}(0) \cap \Omega^*_{U\times\mathbb{R}^n}$ and is not contained in any compact subset of $\Omega^*_{U\times\mathbb{R}^n}$. According to Lemma 7.1, one finds that $\Gamma = \mathfrak{H}(\Theta)$ is a set of T-forced pairs for (7.4) in Ω such that trivial T-triples of (7.13) correspond to trivial T-forced pairs. Since $\mathfrak{H}$ is a homeomorphism, we have the following result:

Theorem 7.4 *Let Ω be an open subset of $[0,\infty)\times C_T(U)$. Let w be as in (7.14) and assume that $\deg(w, \Omega_U)$ is well-defined and nonzero. Then there exists a connected set Γ of nontrivial T-forced pairs in Ω of (7.4) whose closure in $[0,\infty) \times C_T(U)$ intersects the set $\left\{(0, \overline{p}) \in [0,\infty) \times C_T(U) : p \in w^{-1}(0) \cap \Omega_U\right\}$ and is not contained in any compact subset of Ω. In particular, when $U = \mathbb{R}^n$ and $\Omega = [0,\infty) \times C_T(\mathbb{R}^n)$, then Γ is unbounded.*

7.4.2 *On the Set of T-Forced Pairs for Equation (7.5)*

Proceeding analogously to the above case, system (7.7) can be written as

$$\begin{cases} x'(t) = \phi_0^{-1}\big(y(t)\big) + \lambda h(\lambda, x(t), y(t)), \\ y'(t) = g\Big(x(t), \phi_0^{-1}\big(y(t)\big) + \lambda h(\lambda, x(t), y(t))\Big) + \\ \qquad +\lambda f\Big(t, x(t), \phi_0^{-1}\big(y(t)\big) + \lambda h(\lambda, x(t), y(t))\Big). \end{cases} \tag{7.20}$$

With the same procedure used in Remark 7.2, we can isolate λ from inside g as well and get, for some appropriate function $\widehat{h}$, the equality:

$$g\Big(x(t), \phi_0^{-1}\big(y(t)\big) + \lambda h(\lambda, x(t), y(t))\Big) = g\Big(x(t), \phi_0^{-1}\big(y(t)\big)\Big) + \lambda\widehat{h}(\lambda, x(t), y(t)).$$

So, collecting all the λ-dependent terms in the second equation into an appropriate function F, we can rewrite (7.20) as

$$\begin{cases} x'(t) = \phi_0^{-1}\big(y(t)\big) + \lambda h(\lambda, x(t), y(t)), \\ y'(t) = g\Big(x(t), \phi_0^{-1}\big(y(t)\big)\Big) + \lambda F\big(t, x(t), y(t), \lambda\big). \end{cases} \tag{7.21}$$

This is of the form (7.10) with $M := U \times \mathbb{R}^n$, so that, in analogy with Sect. 7.4.1, we will speak about T-triples rather than T-pairs. Accordingly, we will relate the set $\Xi \subseteq [0, \infty) \times C_T^1(U)$ of T-forced pairs of (7.5) with the set $\Upsilon \subseteq [0, \infty) \times C_T(U \times \mathbb{R}^n)$ of T-triples of (7.21).

Define the map $\mathfrak{G}\colon \Upsilon \to \Xi$ by $\mathfrak{H}\colon (\lambda, x, y) \mapsto (\lambda, x)$. A result analogous to Lemma 7.1 holds in this case too.

Lemma 7.3 *$\mathfrak{G}$ is a homeomorphism that respects the notion of triviality, in the sense that it makes trivial T-triples correspond to trivial T-forced pairs and vice versa.*

In this case our result will be a consequence of Theorem 7.1. Let G be the vector field defined on $U \times \mathbb{R}^n$:

$$(p, q) \mapsto \Big(\phi_0^{-1}(q)\ ,\ g\big(p, \phi_0^{-1}(q)\big)\Big).$$

In analogy with the previous section, since ϕ_0^{-1} is a homeomorphism, the degree of G reduces to the degree of the vector field:

$$\gamma(p) := g(p, 0). \tag{7.22}$$

The proof of the next lemma can be carried out as in Lemma 7.2 above and, therefore, is omitted.

Lemma 7.4 *Let $W \subseteq U$ be open. The vector field G is admissible for the degree in $W \times \mathbb{R}^n$ if and only if so is γ in W, and*

$$\deg\big(G, W \times \mathbb{R}^n\big) = \pm \deg\big(\gamma, W\big). \tag{7.23}$$

We conclude as in Sect. 7.4.1. Given an open subset Ω of $[0, \infty) \times C_T(U)$, we observe that the set $\Omega^* = \Omega \times C_T(\mathbb{R}^n)$ is open in $[0, \infty) \times C_T(U \times \mathbb{R}^n)$ and we define the open subset of U:

$$\Omega_U = \big\{p \in U : (0, \overline{p}) \in \Omega\big\}.$$

Furthermore, we consider the open set $\Omega^*_{U\times\mathbb{R}^n} \subseteq U \times \mathbb{R}^n$ given by

$$\Omega^*_{U\times\mathbb{R}^n} = \big\{(p,q) \in U \times \mathbb{R}^n : (0,\overline{p},\overline{q}) \in \Omega^*\big\} = \Omega_U \times \mathbb{R}^n.$$

By Lemma 7.4 it follows that, if $\deg(\gamma, \Omega_U)$ is well-defined and nonzero, then so is $\deg(G, \Omega^*_{U\times\mathbb{R}^n})$. Hence, Theorem 7.1 applied to (7.21) yields the existence of a connected set, say Θ, of nontrivial T-triples in Ω of (7.21) whose closure in $[0,\infty) \times C_T(U \times \mathbb{R}^n)$ meets $G^{-1}(0) \cap \Omega^*_{U\times\mathbb{R}^n}$ and is not contained in any compact subset of $\Omega^*_{U\times\mathbb{R}^n}$. By Lemma 7.3, we get that $\Gamma = \mathfrak{G}(\Theta)$ is a set of T-forced pairs of (7.5) in Ω since $\mathfrak{G}$ is a homeomorphism. As above, observe that a T-triple $(0,x,y)$ of (7.21) is trivial if and only if $(0,x)$ is a trivial T-pair of (7.5). Thus, we have proved the following result:

Theorem 7.5 *Let ω be an open subset of $[0,\infty) \times C_T(U)$. Let γ be as in (7.22) and assume that $\deg\big(\gamma, \Omega_U\big)$ is well-defined and nonzero. Then there exists a connected set Γ of nontrivial T-forced pairs in Ω of (7.5) whose closure in $[0,\infty) \times C_T(U)$ intersects the set $\big\{(0,\overline{p}) \in [0,\infty) \times C_T(U) : p \in \gamma^{-1}(0) \cap \Omega_U\big\}$ and is not contained in any compact subset of Ω. In particular, when $U = \mathbb{R}^n$ and $\Omega = [0,\infty) \times C_T(\mathbb{R}^n)$, then Γ is unbounded.*

As already pointed out, this theorem cannot be obtained as a direct consequence of Theorem 7.4.

7.5 Final Remarks and Perspectives

In this chapter we have investigated the structure of the set of T-periodic solutions of equations (7.1) and (7.2). Although the former can be seen as a particular case of the latter, and despite the fact that the results concerning the set of T-forced pairs of either equations have a similar form, we notice an important difference. Namely, the degree of the average wind w, which is crucial for Theorem 7.4, plays no role in Theorem 7.5: It could, in principle, be not even defined. Conversely, for Eq. (7.1), the degree of γ does not even make sense. It seems natural to compare Theorems 7.2 and 7.1 with Theorems 7.4 and 7.5, respectively.

An interesting question, which we postpone to further research, is whether it would be possible to find a bridge between Theorems 7.4 and 7.5. After all, Theorem 7.3 does something similar for Theorems 7.1 and 7.2, see the pertinent discussion in [35].

A further attractive line of study, not addressed here, is the investigation of the set of T-periodic solutions of Eqs. (7.1) and (7.2) when a dependence on delayed arguments is introduced in ϕ and f. Namely if equations of the form

$$[\phi_\lambda\big(x(t), x(t-r), x'(t)\big)]' = g\big(x(t), x'(t)\big) \\ + \lambda f\big(t, x(t), x(t-r), x'(t), x(t-r)\big), \quad \lambda \geq 0,$$

are considered. In fact, it seems reasonable to combine the method used here to prove Theorem 7.1 with the arguments developed in [27]. Actually, at the price of some increase in the technical details, it seems possible to use the arguments of [14] in conjunction with the techniques of the present paper to further generalize our results to the functional delay case.

Finally, we mention another possible extension of the results obtained here. It seems plausible to use the results of [23, 34] in order to generalize Theorem 7.3 to the case when $\mathfrak{h}$ and $\mathfrak{f}$ are Carathéodory so that, following the argument of Sect. 7.4, one could extend Theorems 7.4 and 7.5 to equations of the form (7.1) and (7.2) where the forcing term f is allowed to be Carathéodory.

References

1. P. Amster, M.P. Kuna, D. Santos, Stability, existence and non-existence of T-periodic solutions of nonlinear delayed differential equations with φ-Laplacian. Commun. Pure Appl. Anal. **21**(8), 2723–2737 (2022)
2. V.I. Arnold, *Ordinary Differential Equations* (Universitext, Springer, Berlin, 2006)
3. F. Battelli, M. Fečkan, General Melnikov approach to implicit ODE's. J. Dyn. Differ. Equations **34**(1), 365–397 (2022)
4. J. Benedikt, P. Girg, L. Kotrla, P. Takáč, Origin of the p-Laplacian and A. Missbach. Electron. J. Differ. Equ. **2018**, Paper No. 16, 17 pp (2018)
5. P. Benevieri, A. Calamai, Bifurcation results for a class of perturbed Fredholm maps. Fixed Point Theory Appl. **2008**, Article ID 752657, 19 pp (2008)
6. C. Bereanu, J. Mawhin, Periodic solutions of nonlinear perturbations of Φ-Laplacians with possibly bounded Φ. Nonlin. Analysis TMA **68**, 1668–1681 (2008)
7. S. Biagi, A. Calamai, F. Papalini, Heteroclinic solutions for a class of boundary value problems associated with singular equations. Nonlin. Analysis TMA **184**, 44–68 (2019)
8. L. Bisconti, A. Calamai, M. Spadini, Periodic solutions of semi-explicit differential-algebraic equations with time-dependent constraints. Bound. Value Probl. **2014**(179), 19 pp (2014)
9. G. Bognár, M. Rontó, Numerical-analytic investigation of the radially symmetric solutions for some nonlinear PDEs. Comput. Math. Appl. **50**(7), 983–991 (2005)
10. A. Boscaggin, G. Feltrin, F. Zanolin, Uniqueness of positive solutions for boundary value problems associated with indefinite Laplacian-type equations. Open Math. **19**, 163–183 (2021)
11. A. Cabada, An overview of the lower and upper solutions method with nonlinear boundary value conditions. Bound. Value Probl. **2011**, Art. ID 893753, 18 pp (2011)
12. A. Calamai, Branches of harmonic solutions for a class of periodic differential-algebraic equations. Commun. Appl. Anal. **15**(2,3 and 4), 273–282 (2011)
13. A. Calamai, Heteroclinic solutions of boundary value problems on the real line involving singular Φ-Laplacian operators. J. Math. Anal. Appl. **378**(2), 667–679 (2011)
14. A. Calamai, M.P. Pera, M. Spadini, Branches of forced oscillations induced by a delayed periodic force. Adv. Nonlinear Stud. **19**(1), 149–163 (2019)
15. A. Calamai, M. Spadini, Periodic perturbations of constrained motion problems on a class of implicitly defined manifolds. Commun. Contemp. Math. **17**(2), 1450027, 19 pp (2015)
16. G. Cupini, C. Marcelli, F. Papalini, Heteroclinic solutions of boundary-value problems on the real line involving general nonlinear differential operators. Differ. Integral Equ. **24**(7–8), 619–644 (2011)
17. G. Dinca, J. Mawhin, Brouwer Degree—the core of nonlinear analysis, in *Progress in Nonlinear Differential Equations and their Applications* (Birkhäuser/Springer, Cham, 2021)

18. N. El Khattabi, M. Frigon, N. Ayyadi, Multiple solutions of boundary value problems with ϕ-Laplacian operators and under a Wintner-Nagumo growth condition. Bound. Value Probl. **2013**, Paper No. 236, 21 pp (2013)
19. C. Fabry, D. Fayyad, Periodic solutions of second order differential equations with a p-Laplacian and asymmetric nonlinearities. Rend. Istit. Mat. Univ. Trieste **24**(1–2), 207–227 (1994)
20. M. Fečkan, A survey on the Melnikov theory for implicit ordinary differential equations with applications to RLC circuits, in *Mathematics Applied to Engineering, Modelling, and Social Issues*, ed. by F.T. Smith, et al. Studies System Decision Control, vol. 200 (Springer, Cham, 2019), pp. 121–160
21. G. Feltrin, E. Sovrano, F. Zanolin, Periodic solutions to parameter-dependent equations with a ϕ-Laplacian type operator. NoDEA, Nonlinear Differ. Equ. Appl. **26**(5), Paper no. 38, 27 pp (2019)
22. G. Feltrin, F. Zanolin, Bound sets for a class of ϕ-Laplacian operators. J. Differ. Equations **297**, 508–535 (2021)
23. M. Furi, M.P. Pera, Carathéodory periodic perturbations of the zero vector field on manifolds. Topol. Methods Nonlinear Anal. **10**(1), 79–92 (1997)
24. M. Furi, M.P. Pera, Global branches of harmonic solutions to periodic ODEs on manifolds. Boll. Un. Mat. Ital. **11-A**, 709–722 (1997)
25. M. Furi, M.P. Pera, M. Spadini, A set of axioms for the degree of a tangent vector field on differentiable manifolds. Fixed Point Theory Appl. **2010**, Art. ID 845631, 11 pp (2010)
26. M. Furi, M. Spadini, On the set of harmonic solutions to periodically perturbed autonomous differential equations on manifolds. Nonlin. Analysis TMA **29**, 963–470 (1997)
27. M. Furi, M. Spadini, Periodic perturbations with delay of autonomous differential equations on manifolds. Adv. Nonlinear Stud. **9**(2), 263–276 (2009)
28. V. Guillemin, A. Pollack, *Differential Topology* (Prentice-Hall Inc., Englewood Cliffs, 1974)
29. M.W. Hirsch, Differential topology, in *Graduate Texts in Mathematics*, vol. 33 (Springer, Berlin, 1976)
30. J.W. Milnor, *Topology from the Differentiable Viewpoint* (University Press of Virginia, Charlottesville, 1965)
31. L. Nirenberg, *Topics in Nonlinear Functional Analysis* (Courant Institute of Mathematical Sciences, New York, 1974)
32. M. Picasso, J. Rappaz, A. Reist, M. Funk, H. Blatter, Numerical simulation of the motion of a two-dimensional glacier. Int. J. Numer. Methods Eng. **60**(5), 995–1009 (2004)
33. I. Rachůnková, M. Tvrdý, Periodic problems with ϕ-Laplacian involving non-ordered lower and upper functions. Fixed Point Theory **6**(1), 99–112 (2005)
34. M. Spadini, Harmonic solutions of periodic Carathéodory perturbations of autonomous ODE's on manifolds. Nonlin. Analysis TMA **41A**, 477–487 (2000)
35. M. Spadini, Branches of harmonic solutions to periodically perturbed coupled differential equations on manifolds. Discrete Contin. Dyn. Syst. **15**(3), 951–964 (2006)
36. M. Spadini, A note on topological methods for a class of differential-algebraic equations. Nonlin. Analysis TMA **73**(4), 1065–1076 (2010)

Chapter 8
Atypical Bifurcation for a Class of Delay Differential Equations

Pierluigi Benevieri

8.1 Introduction

In this chapter we study a class of neutral delay differential equations, depending on a real parameter λ and represented in the following form:

$$u'(t) + \lambda\phi(t, u(t), u(t-\tau), u'(t)) + \lambda\psi(t, u(t), u(t-\tau), u'(t)) = 0. \tag{8.1}$$

Here, $\phi, \psi : \mathbb{R} \times \mathbb{R}^n \times \mathbb{R}^n \times \mathbb{R}^n \to \mathbb{R}^n$ are assumed to be, respectively, C^1 and continuous and both T-periodic with respect to the first variable. The positive constant time lag τ and the positive period T are given. A solution is defined in the classical sense, that is, as a C^1 map that satisfies the equality in $\mathbb{R}$ or in a smaller interval I.

The study of Eq. (8.1) is inspired by an analogous, undelayed problem, tackled in [3]. In that work, the following equation

$$x'(t) + \lambda\phi(t, x(t), x'(t)) + \lambda\psi(t, x(t), x'(t)) = 0 \tag{8.2}$$

is studied, and a global bifurcation result for periodic solutions is obtained. In this chapter, we proceed in the general spirit of [3], obtaining a similar result for problem (8.1).

Problem (8.1) can be seen as a particular case of a nonlinear operator equation

$$Lu + \lambda(\Phi(u) + \Psi(u)) = 0 \tag{8.3}$$

P. Benevieri (✉)
Instituto de Matemática e Estatística, Universidade de São Paulo, São Paulo, São Paulo, Brazil
e-mail: pluigi@ime.usp.br

P. Amster, P. Benevieri (eds.), *Topological Methods for Delay and Ordinary Differential Equations*, Advances in Mechanics and Mathematics 51,
https://doi.org/10.1007/978-3-031-61337-1_8

in $\mathbb{R} \times \Omega$, where Ω is an open subset of a real Banach space E, and $L : E \to F$ is a linear Fredholm operator of index zero, being F another real Banach space. The maps $\Phi, \Psi : \Omega \to F$ are of class C^1 and continuous, respectively.

The *solutions* of (8.3) are those pairs (λ, u) such that u solves the equation with the corresponding λ. If $\lambda = 0$, the solution is called *trivial*, and the set of the trivial solutions coincides with $\{0\} \times (\Omega \cap \operatorname{Ker} L)$, which we suppose not empty.

Two bifurcation problems can be considered: The first one, called *local*, concerns under what conditions the set of nontrivial solutions is not empty. The second one is usually called *global* and deals with the topological properties of the set of solutions.

A *bifurcation point* is defined as a point p in $\Omega \cap \operatorname{Ker} L$ such that $(0, p)$ lies in the closure of the set of nontrivial solutions. This kind of bifurcation point is sometimes called *atypical*, see, e.g., [25, Section 19], to distinguish problems as (8.3) above from equations $f(\lambda, x) = 0$ in which $f(\lambda, 0) = 0$ for all $\lambda \in \mathbb{R}$, being $\mathbb{R} \times \{0\}$ the set of the trivial solutions. This is the case, for example, of the seminal results by Krasnoselskii [21] and Rabinowitz [26] (see also [16]).

Under additional assumptions, we will specify later on, and in this chapter we obtain, as our main result (Theorem 8.4), a global bifurcation result for T-periodic solutions of problem (8.1). More precisely, we show the existence of a connected (in a suitable topology) branch of nontrivial solutions of (8.1) whose closure is not compact and contains a trivial solution. Our approach is topological, and it is based on a concept of oriented degree for noncompact perturbations of Fredholm maps of index zero between Banach spaces, introduced in [1, 2].

This chapter is divided as follows. The second section deals with some basic concepts concerning the measure of noncompactness associated to maps. The third section is devoted to recalling some concepts of orientability and orientation for Fredholm maps of index zero between real Banach spaces. In the fourth section, we bring the concepts and important results concerning orientation and degree for a special class of noncompact perturbations of Fredholm maps. The fifth section deals with some results concerning the atypical bifurcation. In the sixth section, we present our main results concerning global bifurcation for periodic solutions of (8.1). In the last section we apply the bifurcation results to equations of Mackey–Glass type and Duffing type that are important, among other applications, in continuum mechanics. Sections 8.2–8.5 concern preliminary results, proven in previous papers. The space we dedicate in this chapter to the preliminaries is certainly unusual, but we find it useful for the reader's understanding of the principal results of the last two sections.

8.2 Preliminaries: Measures of Noncompactness

We recall in this section the definition and the basic properties of the α-Fredholm maps, a special class of maps, introduced in [1]. In this section, E and F will always stand for two real infinite-dimensional Banach spaces. Given a bounded subset A of E, the *Kuratowski measure of noncompactness* $\alpha(A)$ of A is the infimum of those positive real numbers d such that A admits a finite covering by sets of diameter less than d. If A is unbounded, we set $\alpha(A) = +\infty$. See, e.g., [15, 16, 22] for a general presentation and properties of this concept.

Given an open subset Ω of E and a continuous map $f : \Omega \to F$, the following two extended real numbers can be associated with f:

$$\alpha(f) = \sup\left\{\frac{\alpha(f(A))}{\alpha(A)} : A \subseteq \Omega \text{ bounded}, \alpha(A) > 0\right\},$$

$$\omega(f) = \inf\left\{\frac{\alpha(f(A))}{\alpha(A)} : A \subseteq \Omega \text{ bounded}, \alpha(A) > 0\right\}.$$

For a short list of properties of $\alpha(f)$ and $\omega(f)$ the reader can see [18]. We just mention here that $\alpha(f) = 0$ if and only if f is completely continuous and $\omega(f) > 0$ only if f is proper on bounded closed sets. The following proposition states an important property for the linear operators (see [18, Proposition 3.2.1]).

Proposition 8.1 *Let $L : E \to F$ be a bounded linear operator. Then $\omega(L) > 0$ if and only if* $\operatorname{Im} L$ *is closed and* $\dim \operatorname{Ker} L < +\infty$.

Let $p \in \Omega$ be fixed. The following definitions of $\alpha_p(f)$ and $\omega_p(f)$ have been introduced by A. Calamai in [13] (see also [14]). They can be viewed as the localization of the above $\alpha(f)$ and $\omega(f)$. Denoting by $B(p, s)$ the open ball in E centered at p with radius $s > 0$, the function

$$s \mapsto \alpha(f|_{B(p,s)}) = \sup\left\{\frac{\alpha(f(A))}{\alpha(A)} : A \subseteq B(p, s),\ \alpha(A) > 0\right\}$$

is well defined and nondecreasing in any interval $(0, \delta]$ such that $B(p, \delta) \subseteq \Omega$. Therefore, we can define

$$\alpha_p(f) = \lim_{s\to 0} \alpha(f|_{B(p,s)}).$$

Analogously, one can define $\omega(f|_{B(p,s)})$ and

$$\omega_p(f) = \lim_{s\to 0} \omega(f|_{B(p,s)}).$$

Observe that $\alpha_p(f) \leq \alpha(f)$ and $\omega_p(f) \geq \omega(f)$. The following proposition has been proven in [13].

Proposition 8.2 *Let $f : \Omega \to F$ be continuous and $p \in \Omega$. Then:*

1. $\alpha_p(f) - \alpha_p(g) \leq \alpha_p(f + g) \leq \alpha_p(f) + \alpha_p(g)$.
2. $\omega_p(f) - \alpha_p(g) \leq \omega_p(f + g) \leq \omega_p(f) + \alpha_p(g)$.
3. *If f is locally compact, $\alpha_p(f) = 0$.*

4. *If* $\omega_p(f) > 0$, f *is locally proper at* p.
5. *If* f *is a local homeomorphism and* $\omega_p(f) > 0$, *then* $\alpha_q(f^{-1})\omega_p(f) = 1$, *where* $q = f(p)$.

For bounded linear operators the local numbers α_p and ω_p coincide with α and ω. For C^1 maps, we have the following result (see [13] for the proof). In the statement, $f'(p)$ denotes the Fréchet derivative.

Proposition 8.3 *Let* $f : \Omega \to F$ *be a* C^1 *map. Then, for any* $p \in \Omega$ *we have* $\alpha_p(f) = \alpha(f'(p))$ *and* $\omega_p(f) = \omega(f'(p))$.

If $f : \Omega \to F$ is a Fredholm map, as a straightforward consequence of Propositions 8.1 and 8.3, we obtain $\omega_p(f) > 0$ for any $p \in \Omega$.

The following proposition states useful results for a direct computation of α and ω of bounded linear operators and nonlinear maps. See [3] for the proof.

Proposition 8.4 *Let* $P : E \to E$ *and* $Q : F \to F$ *be two projections onto finite-dimensional subspaces:*

1. *If* $L : E \to F$ *is a bounded linear operator, then*

$$\alpha(L) = \alpha((I - Q)L(I - P)) \quad \textit{and} \quad \omega(L) = \omega((I - Q)L(I - P)).$$

2. *If* $f : \Omega \to F$ *is continuous and* $p \in \Omega$ *is given, then*

$$\alpha_p(f) = \alpha_p((I - Q)f) \quad \textit{and} \quad \omega_p(f) = \omega_p((I - Q)f).$$

Next result, whose proof can be found in [2, Proposition 4.5], extends to the continuous case an analogous result shown in [13] for C^1 maps.

Proposition 8.5 *Let* $g : \Omega \to F$ *and* $\sigma : \Omega \to \mathbb{R}$ *be continuous. Consider the product map* $f : \Omega \to F$ *defined by* $f(x) = \sigma(x)g(x)$. *Then, for any* $p \in \Omega$ *we have* $\alpha_p(f) = |\sigma(p)|\alpha_p(g)$ *and* $\omega_p(f) = |\sigma(p)|\omega_p(g)$.

In the sequel we will also consider maps G defined on the product space $\mathbb{R} \times E$. In order to define $\alpha_{(\lambda,p)}(G)$, we consider the norm

$$||(\lambda, p)|| = \max\{|\lambda|, \|p\|\}.$$

Remark 8.1 The natural projection of $\mathbb{R} \times E$ onto E, denoted here by π_2, is nonexpansive. Therefore $\alpha(\pi_2(X)) \leq \alpha(X)$ for any subset X of $\mathbb{R} \times E$. More precisely, since $\mathbb{R}$ is finite dimensional, if $X \subseteq \mathbb{R} \times E$ is bounded, we have $\alpha(\pi_2(X)) = \alpha(X)$.

We conclude the section with the following technical result, which is a straightforward consequence of Proposition 8.5.

Corollary 8.1 *Given a continuous map* $\varphi : \Omega \to F$, *consider the map*

$$\Phi : [0, 1] \times \Omega \to F, \qquad \Phi(\lambda, x) = \lambda\varphi(x).$$

Then, for any fixed pair $(\lambda, p) \in [0, 1] \times \Omega$, *we have*

$$\alpha_{(\lambda,p)}(\Phi) = \lambda\alpha_p(\varphi).$$

8.3 A Concept of Orientation for Linear Fredholm Operators and Nonlinear Fredholm Maps

In this section, we recall the concepts, introduced in [4, 5], of orientability and orientation for linear Fredholm operators and nonlinear Fredholm maps of index zero between real Banach spaces. In this section E and F will stand for real Banach spaces of not necessarily infinite dimension. Consider a bounded Fredholm linear operator of index zero $L : E \to F$, also called Φ_0-*operator* in the sequel. A bounded linear operator $A : E \to F$ is said to be a *corrector* of L if:

(i) The image of A is finite dimensional.
(ii) $L + A$ is an isomorphism.

On the set $C(T)$ of correctors of T, which is nonempty as it is easy to see, one can define an equivalence relation as follows. Let $A, B \in C(L)$ be given and consider the following automorphism of E:

$$T = (L + B)^{-1}(L + A) = I - (L + B)^{-1}(B - A).$$

Consider a nontrivial finite-dimensional subspace E_0 of E containing the image of $(L + B)^{-1}(B - A)$. T maps E_0 into itself, and, consequently, the determinant of its restriction $T_0 : E_0 \to E_0$ is well defined and does not depend on the choice of E_0. Thus, the *determinant* of T, $\det(T)$, is well defined as the determinant of the restriction of T to any nontrivial finite-dimensional subspace of E containing the image of $(L + B)^{-1}(B - A)$ (see, e.g., [20, § III-4] and references therein). We say that A is *equivalent* to B if

$$\det\Big((L + B)^{-1}(L + A)\Big) > 0.$$

As shown in [4], this is actually an equivalence relation in $C(L)$, with two equivalence classes, and it provides the following concept of orientation for bounded Fredholm linear operators of index zero (see [4]).

Definition 8.1 (Orientation of a Φ_0-Operator) An *orientation* of a Φ_0-operator $L : E \to F$ is the choice of one of the two equivalence classes of $C(L)$. When an orientation is chosen, L is called *oriented*, and the elements of the orientation are called *positive correctors* of L.

Denote by $\mathcal{L}(E, F)$ the Banach space of the bounded linear operators of E into F and by $\Phi_0(E, F)$ the open subset of $\mathcal{L}(E, F)$ of the Φ_0-operators. Consider $L \in \Phi_0(E, F)$, $A \in C(L)$, and suppose L be oriented with A positive corrector. Since the set of the isomorphisms of E into F is open in $\mathcal{L}(E, F)$, A turns out to be a corrector of every T in a suitable neighborhood W of L in $\Phi_0(E, F)$. Thus, any $T \in W$ can be oriented by taking A as a positive corrector. This fact allows to give a definition of orientation for nonlinear Fredholm maps. Recall that, given an open subset Ω of E, $g\colon \Omega \to F$ is a *Fredholm map* if it is C^1 and its Fréchet derivative, $g'(x)$, is a Fredholm operator for all $x \in \Omega$. The *index* of g is n if $g'(x)$ has index n for all x of the domain. We will call Φ_0-*map* a nonlinear Fredholm map of index zero.

Definition 8.2 (Orientation of a Φ_0-map) An *orientation* of a Φ_0-map $g\colon \Omega \to F$ is the choice of an orientation $\alpha(x)$ of the derivative $g'(x)$ for every $x \in \Omega$ such that the following condition holds: For any $x \in \Omega$ there exists $A \in \alpha(x)$ that belongs to $\alpha(y)$ for any y in a neighborhood of x. A map g is *orientable* if it admits an orientation and is *oriented* when an orientation is chosen.

A very special case of an orientable map is a Φ_0-operator $T : E \to F$. In this case, T, being also a Φ_0-map, can be oriented according to Definition 8.1 or to the 8.2. Hereafter, we will implicitly assume that the two notions agree in the sense that, if T is oriented according to Definition 8.1, then the differential of T at any $x \in E$, which clearly coincides with T, assumes the same orientation and vice versa.

We conclude the section with the following important notion of orientation for a homotopy of Φ_0-maps.

Definition 8.3 Let Ω be open in E. We say that a C^1 map $H\colon [0,1] \times \Omega \to F$ is a Φ_0-*homotopy* if any partial map $H_\lambda := H(\lambda, \cdot)$ is a Φ_0-map. An *orientation* of $H\colon [0,1] \times \Omega \to F$ is an orientation of the partial derivative map

$$\partial_1 H\colon [0,1] \times \Omega \to \Phi_0(E, F), \quad (\lambda, x) \mapsto (H_\lambda)'(x),$$

according to Definition 8.2. Moreover, H is *orientable*, or *oriented*, if so is $\partial_1 H$.

Notice that, if a Φ_0-homotopy $H\colon [0,1] \times \Omega \to F$ is oriented, any partial map $H_\lambda\colon \Omega \to F$ inherits an orientation induced by H. The following theorem states that the converse is true.

Theorem 8.1 *Let $H\colon [0,1] \times \Omega \to F$ be a Φ_0-homotopy. Given $\lambda \in [0,1]$, assume that the partial map H_λ is oriented and call α_λ its orientation. Then:*

(i) Any partial map H_μ is orientable.
(ii) There exists and is unique an orientation of H which induces the orientation α_λ on H_λ.

8.4 Degree for α-Fredholm Triples

In this section we summarize the construction of a topological degree introduced in [1, 2] (see also the excellent book [28] by M. Väth) for a special class of noncompact perturbations of Φ_0-maps, called *α-Fredholm maps*, whose definitions, which are given in terms of the numbers α_p and ω_p, have been recalled in Sect. 8.2. More precisely, an *α-Fredholm map* $f : \Omega \to F$ is of the form $f = g - k$, where g is a Φ_0-map, k is continuous, and $\alpha_p(k) < \omega_p(g)$ for every p. The degree is given as an integer-valued map defined on a class of triples that we shall call *admissible α-Fredholm triples*. This class is recalled in the following definition.

Definition 8.4 Consider two real Banach spaces E and F and an open subset Ω of E. Let $g : \Omega \to F$ be a Φ_0-map of index zero, $k : \Omega \to F$ a continuous map, and U an open subset of Ω. The triple (g, U, k) is said to be an *α-Fredholm triple* if for any $p \in U$ we have

$$\alpha_p(k) < \omega_p(g).$$

An α-Fredholm triple (g, U, k) is said to be *admissible* for the degree if:

(i) g is oriented according to Definition 8.2.
(ii) The *coincidence set* $S = \{x \in U : g(x) = k(x)\}$ is compact.

The definition of a topological degree for the admissible α-Fredholm triples is the result of a complex process we developed in [1, 2] (see also [6, 7]). Here, we only give a simplified summary. We divide the construction in three steps.

Step 1 Let (g, U, k) be an admissible α-Fredholm triple and suppose that $k(U)$ is contained in a finite-dimensional subspace of F. Denote $f := g - k$. As $f^{-1}(0) \cap U$ is compact, there exist a finite-dimensional subspace Z of F and an open neighborhood $W \subseteq U$ of $f^{-1}(0) \cap U$ such that g is transverse to Z in W. Assume that Z contains $k(U)$. Chosen any orientation of Z, the C^1 manifold $M = g^{-1}(Z) \cap W$ can be oriented with an orientation induced by the orientations of g and Z (see [6]). Let $f|_M$ denote the restriction of f to M, as domain, and to Z, as codomain. One can easily verify that $(f|_M)^{-1}(0) = f^{-1}(0) \cap U$. Thus $(f|_M)^{-1}(0)$ is compact, and the Brouwer degree $\deg_B(f|_M, M, 0)$ is consequently well defined. The *degree* of the admissible triple (g, U, k) is the integer

$$\deg(g, U, k) = \deg_B(f|_M, M, 0). \tag{8.4}$$

In [6] it is shown that this definition is well posed since the right-hand side of (8.4) is independent of the choice of W and Z.

Step 2 Let us now extend the definition of degree to admissible α-Fredholm triples (g, U, k) such that k is locally compact. As in step 1, denote $f := g - k$. Consider:

1. An open neighborhood V of $f^{-1}(0) \cap U$ such that $\overline{V} \subseteq U$, g is proper on $\overline{V}$, and $k|_{\overline{V}}$ has relatively compact image.
2. A continuous map $\xi : \overline{V} \to F$ having bounded finite-dimensional image and such that

$$\|k(x) - \xi(x)\| < \rho, \quad \forall x \in \partial V,$$

where ρ is the distance in F between 0 and the set $f(\partial V)$, which is closed, since f is proper on $\overline{V}$, being the sum of the proper map $g|_{\overline{V}}$ and the compact map $k|_{\overline{V}}$.

Then, define

$$\deg(g, U, k) = \deg(g, V, \xi), \tag{8.5}$$

where the right-hand side is defined in formula (8.4). In [6], it is shown that the above definition is well posed since formula (8.5) does not depend on g, ξ, and V.

Step 3 Let (g, U, k) be an admissible α-Fredholm triple. Given a finite covering $\mathcal{V} = \{V_1, \ldots, V_N\}$ of open balls of S and a compact convex set C, with $S \subseteq C \subseteq U$, the pair $(\mathcal{V}, C)$ is called an *α-pair* relative to (g, U, k) if, for any $i = 1, \ldots, N$, the following conditions hold:

(1) The ball $\widetilde{V}_i$ of double radius and same center as V_i is contained in U.
(2) $\alpha(k|_{\widetilde{V}_i}) < \omega(g|_{\widetilde{V}_i})$.
(3) $\{x \in V_i : g(x) \in k(\widetilde{V}_i \cap C)\} \subseteq C$.

In [1] it is shown that, given any admissible α-Fredholm triple, it is always possible to find a relative α-pair. Let $(\mathcal{V}, C)$ be an α-pair relative to (g, U, k). Denote $V = \bigcup_{i=1}^{N} V_i$ and consider a retraction $r : E \to C$, whose existence is ensured by Dugundji's Extension Theorem (see, e.g., [17]). Let W be an open subset of V containing S such that, for any i, $x \in W \cap V_i$ implies $r(x) \in \widetilde{V}_i$. Notice that the triple (g, W, kr) is as in step 2, and thus the degree of the (g, U, k) is

$$\deg(g, U, k) = \deg(g, W, kr).$$

It is shown, in [1], that the right-hand side of the above equality is independent of the choice of the α-pair $(\mathcal{V}, C)$, of the retraction r and of the open set W.

As pointed out in [1], this concept of degree extends the Leray–Schauder degree, and it agrees with the Nussbaum degree for the class of locally α-contractive vector fields (see [24]).

Below we state the most important properties of the degree. The reader can see [1] for the relative proofs. We need first the following concept of *α-Fredholm homotopy*.

Definition 8.5 Let W be an open subset of $[0, 1]\times E$ and $H : W \to F$ a continuous map of the form $H(\lambda, x) = G(\lambda, x) - K(\lambda, x)$. We say that H is an α-*Fredholm homotopy* if the following conditions hold:

(i) G is C^1 and, for any $\lambda \in [0, 1]$, the partial map G_λ is Fredholm of index zero on the section $W_\lambda = \{x \in E : (\lambda, x) \in W\}$.
(ii) For any pair $(\lambda, p) \in W$ we have $\alpha_{(\lambda,p)}(K) < \omega_{(\lambda,p)}(G)$.

Theorem 8.2 *The degree satisfies the following properties:*

1. *(Normalization) Let the identity I of E be oriented in such a way that the trivial operator is a positive corrector. Then*

$$\deg(I, E, 0) = 1.$$

2. *(Additivity) Given an admissible α-Fredholm triple (g, U, k) and two disjoint open subsets U_1, U_2 of U, assume that $S = \{x \in U : g(x) = k(x)\}$ is contained in $U_1 \cup U_2$. Then*

$$\deg(g, U, k) = \deg(g, U_1, k) + \deg(g, U_2, k).$$

3. *(Excision) Given an admissible α-Fredholm triple (g, U, k) and an open subset U_1 of U, assume that S is contained in U_1. Then*

$$\deg(g, U, k) = \deg(g, U_1, k).$$

4. *(Existence) Given an admissible α-Fredholm triple (g, U, k), if*

$$\deg(g, U, k) \neq 0,$$

then the equation $g(x) = k(x)$ has a solution in U.

5. *(Homotopy invariance) Let W be an open subset of $[0, 1] \times E$ and $H : W \to F$ an α-Fredholm homotopy of the form $H(\lambda, x) = G(\lambda, x) - K(\lambda, x)$. Assume that G is oriented and that the set $H^{-1}(0)$ is compact. Then $\deg(G_\lambda, W_\lambda, K_\lambda)$ is well defined and does not depend on $\lambda \in [0, 1]$.*

8.5 Atypical Bifurcation Results

In this section we summarize the bifurcation results obtained in [3] concerning the semilinear operator equation

$$Lx + \lambda(h(x) + k(x)) = 0. \tag{8.6}$$

Without loss of generality, assume $\lambda \geq 0$. Here, $L : E \to F$ is a linear Fredholm operator of index zero between real Banach spaces, and the maps $h : \Omega \to F$ and $k : \Omega \to F$ are, respectively, C^1 and continuous in an open subset Ω of E.

Equation (8.6) can be equivalently written as

$$H(\lambda, x) = 0,$$

where

$$H : [0, +\infty) \times \Omega \to F, \quad H(\lambda, x) = Lx + \lambda(h(x) + k(x)).$$

This map can also be represented as

$$H(\lambda, x) = G(\lambda, x) - K(\lambda, x), \tag{8.7}$$

where $G(\lambda, x) = Lx + \lambda h(x)$ and $K(\lambda, x) = -\lambda k(x)$. We will suppose that the following conditions hold:

(*H1*) For any $\lambda \geq 0$ the partial map G_λ is Fredholm of index zero.

(*H2*) For any pair $(p, \lambda) \in [0, +\infty) \times \Omega$ we have $\alpha_{(\lambda,p)}(K) < \omega_{(p,\lambda)}(G)$.

Thus, the map H is an α-Fredholm homotopy (see Definition 8.5). By a solution of (8.6) we mean a pair $(\lambda, x) \in H^{-1}(0)$, and we regard the subset $\{0\} \times (\Omega \cap \operatorname{Ker} L)$ of $H^{-1}(0)$ as the set of *trivial solutions*. A problem related to Eq. (8.6) concerns the existence of an *atypical bifurcation point* (in the terminology of Prodi–Ambrosetti in [25]), that is, a point p in $\Omega \cap \operatorname{Ker} L$ such that $(0, p)$ lies in the closure of the set of nontrivial solutions.

In [8] it is obtained a global bifurcation result for Eq. (8.6) in the particular case when $k = 0$. Afterward, the result in [8] was extended in [3, 6] by introducing a perturbation k. We also cite the recent paper [9] in which an analogous atypical global bifurcation problem is tackled for dynamic equations in time scales. Theorem 8.3 has been proven in [3] for a not necessarily compact perturbation k, where the compactness assumption of k is replaced by condition *H2)* above, which is clearly satisfied when k (and thus K) is locally compact.

Before recalling Theorem 8.3, we need some terminology. Let F_1 be any finite-dimensional direct complement of $\operatorname{Im} L$ in F. Consider the decomposition $F = \operatorname{Im} L \oplus F_1$, and denote by R and π the associated projections onto $\operatorname{Im} L$ and F_1, respectively. Equation (8.6) is clearly equivalent to the system

$$\begin{cases} Lx + \lambda(Rh(x) + Rk(x)) = 0 \\ \lambda(\pi h(x) + \pi k(x)) = 0. \end{cases} \tag{8.8}$$

In order to investigate the set of nontrivial solutions of (8.8), it is convenient to consider the system

$$\begin{cases} Lx + \lambda(Rh(x) + Rk(x)) = 0 \\ \pi h(x) + \pi k(x) = 0, \end{cases} \tag{8.9}$$

which is equivalent to (8.8) for $\lambda \neq 0$. The fundamental tools for proving Theorem 8.3 are the homotopy invariance property of the degree for α-Fredholm maps (as in Theorem 8.2), together with the following Whyburn-type lemma, whose proof can be found in [19].

Lemma 8.1 *Let Z be a compact subset of a locally compact metric space X. Assume that any compact subset of X containing Z has nonempty boundary. Then $X \setminus Z$ contains a connected set whose closure is not compact and intersects Z.*

Theorem 8.3 *Let $H : [0, +\infty) \times \Omega \to F$ be defined by $H(\lambda, x) = Lx + \lambda(h(x) + k(x))$ and suppose that conditions (H1) and (H2) above hold. Assume that the map G of (8.7) is oriented. Let $v : \Omega \cap \operatorname{Ker} L \to F_1$ be defined by $v(p) = \pi h(p) + \pi k(p)$. Let W be an open subset of $[0, +\infty) \times \Omega$ and suppose that the Brouwer degree $\deg_B(v, W_0 \cap \operatorname{Ker} L, 0)$ is well defined and nonzero. Then, there exists in W a connected set of nontrivial solutions of (8.6) whose closure in W is not compact and intersects $\{0\} \times \operatorname{Ker} L$.*

This theorem has been presented and proven in [3]. Its proof goes beyond the scope of this chapter. However, we find useful to summarize here the idea in order to show the importance of the degree for the α-Fredholm maps, not appearing in the statement.

Sketch of the Proof Define $\widehat{H} : W \to F = \operatorname{Im} L \oplus F_1$ by

$$\widehat{H}(\lambda, x) = Lx + \lambda(R\Phi(x) + R\Psi(x)) + \pi(\Phi(x) + \Psi(x)).$$

It is not difficult to see that $\widehat{H}$ is an oriented α-Fredholm homotopy, by the decomposition

$$\widehat{H}(\lambda, x) = \widehat{\Phi}(\lambda, x) - \widehat{\Psi}(\lambda, x),$$

where

$$\widehat{\Phi}(\lambda, x) = Lx + \lambda R\Phi(x) + \pi\Phi(x), \quad \text{and} \quad \widehat{\Psi}(\lambda, x) = -\lambda R\Psi(x) - \pi\Psi(x).$$

The set

$$A = \left\{(\lambda, x) \in W : \widehat{H}(\lambda, x) = 0\right\}$$

is locally compact since the map $\widehat{H}$ is locally proper. Moreover, $A_0 = v^{-1}(0) \cap W_0$ is compact because we assumed that $\deg_B(v, W_0 \cap \operatorname{Ker} L, 0)$ is well defined.

We need to verify that any compact subset of A containing $A_0 \times \{0\}$ has nonempty boundary in A, and, to this purpose, it is crucial the use of the degree for the α-Fredholm maps. Assume the contrary and call A' a compact subset of A containing $\{0\} \times A_0$, with empty boundary in A. Hence, A' is an open subset of A, and there exists a bounded open subset U of W such that $A' = U \cap Y$. Since A' is compact, the homotopy invariance property of the degree implies that $\deg(\widehat{\Phi}_\lambda, U_\lambda, \widehat{\Psi}_\lambda)$ does not depend on $\lambda \geq 0$. Moreover, the slice $A'_{\lambda'} = U_{\lambda'} \cap A_{\lambda'}$ is empty for some positive λ'. By the existence property of the degree (Theorem 8.2), we have that

$$\deg(\widehat{\Phi}_{\lambda'}, U_{\lambda'}, \widehat{\Psi}_{\lambda'}) = 0.$$

Hence, $\deg(\widehat{\Phi}_\lambda, U_\lambda, \widehat{\Psi}_\lambda) = 0$ for any $\lambda \in [0, +\infty)$, by the homotopy invariance property. In particular,

$$\deg(\widehat{\Phi}_0, U_0, \widehat{\Psi}_0) = 0. \tag{8.10}$$

Using the excision property of the degree and a finite-dimensional reduction argument (analogous to that in the construction of the Leray–Schauder degree, see [3]), we finally obtain

$$\deg_B(v, W_0 \cap \operatorname{Ker} L, 0) = 0,$$

which contradicts the assumption for the above degree in the statement of the theorem. Thus, we can apply the above Whyburn-type result, Lemma 8.1, and conclude that there exists a connected subset of A whose closure in A intersects $\{0\} \times A_0$ and is not compact. This is the idea of the proof. □

Some consequences of Theorem 8.3, Corollaries 8.2 and 8.3 extend analogous results in [8]. See [3] for the proofs.

Corollary 8.2 *Let the assumptions of Theorem 8.3 be satisfied. Suppose, moreover, that the map H is proper on bounded and closed subsets of W. Then, Eq.* (8.6) *admits a connected set Γ of nontrivial solutions such that its closure in* $[0, +\infty) \times E$ *intersects* $\{0\} \times \operatorname{Ker} L$ *and is either unbounded or reaches the boundary of W. If, in particular, $\Omega = E$ and $W = [0, +\infty) \times E$, then Γ is unbounded.*

Corollary 8.3 *Let W and v be as in Theorem 8.3. Suppose, moreover, that the map H is proper on bounded and closed subsets of W. Let $p \in W_0 \cap \operatorname{Ker} L$ be such that $v(p) = 0$ and $v'(p)$ is invertible. Then, Eq.* (8.6) *admits a connected set Γ of nontrivial solutions such that its closure contains p and satisfies at least one of the following three conditions:*

(i) It is unbounded.
(ii) It contains a point $q \in W_0 \cap \operatorname{Ker} L$, $q \neq p$.
(iii) It intersects ∂W.

8.6 Global Bifurcation for Nonlinear Parameterized Delay Equations

In this section we come back to Eq. (8.1), and we present our global bifurcation result, Theorem 8.4, based on the general theory we summarized in the previous sections. We will denote by C_T^0 the Banach space of continuous and T-periodic functions of $\mathbb{R}$ into $\mathbb{R}^n$ with the usual supremum norm

$$\|u\|_\infty = \max_{t\in\mathbb{R}} |u(t)|,$$

where $|\cdot|$ denotes the Euclidean norm in $\mathbb{R}^n$. By C_T^1 we will call the space of C^1 and T-periodic functions of $\mathbb{R}$ into $\mathbb{R}^n$, endowed with the norm

$$\|u\|_1 = \|u\|_\infty + \|u'\|_\infty.$$

We define

$$\begin{aligned} \Phi &: C_T^1 \to C_T^0, \quad \Phi(u)(t) = \phi(t, u(t), u(t-\tau), u'(t)), \\ \Psi &: C_T^1 \to C_T^0, \quad \Psi(u)(t) = \psi(t, u(t), u(t-\tau), u'(t)), \end{aligned}$$

and set

$$\begin{aligned} \widetilde{\Phi} &: [0, +\infty) \times C_T^1 \to C_T^0, \quad \widetilde{\Phi}(\lambda, u) = u' + \lambda\Phi(u), \\ \widetilde{\Psi} &: [0, +\infty) \times C_T^1 \to C_T^0, \quad \widetilde{\Psi}(\lambda, u) = -\lambda\Psi(u). \end{aligned} \tag{8.11}$$

$\widetilde{\Phi}$ is C^1, since so is ϕ, and the Fréchet derivative $(\widetilde{\Phi}_\lambda)'(u) : C_T^1 \to C_T^0$ of any partial map $\widetilde{\Phi}_\lambda(\cdot) := \widetilde{\Phi}(\lambda, \cdot)$ at any $u \in C_T^1$ is the bounded linear operator defined by

$$\begin{aligned} (\widetilde{\Phi}_\lambda)'(u)q(t) &= q'(t) + \lambda\partial_2\phi(t, u(t), u(t-\tau), u'(t))q(t) \\ &\quad + \lambda\partial_3\phi(t, u(t), u(t-\tau), u'(t))q(t-\tau) \\ &\quad + \lambda\partial_4\phi(t, u(t), u(t-\tau), u'(t))q'(t), \qquad q \in C_T^1. \end{aligned}$$

Here, any $\partial_i\phi$, $i = 2, 3, 4$, denotes the partial derivative of ϕ with respect to the corresponding variable. Now, for any given λ and u, consider the linear operator

$$G_{(\lambda,u)} : C_T^1 \to C_T^0, \qquad G_{(\lambda,u)}q(t) = q'(t) + \phi(t, u(t), u(t-\tau), u'(t))q'(t), \tag{8.12}$$

which can be written, in a different notation, as

$$G_{(\lambda,u)}q(t) = (I + \lambda M_u(t))q'(t),$$

where M_u is an $n \times n$ matrix, depending on u, of continuous T-periodic real functions defined in $\mathbb{R}$. A necessary and sufficient condition for $G_{(\lambda,u)}$ to be Fredholm of index zero is that

$$\det(I + \lambda M_u(t)) \neq 0, \qquad \text{for any } t \in \mathbb{R}. \tag{8.13}$$

Indeed, if condition (8.13) is satisfied, then the kernel of $G_{(\lambda,u)}$ is n dimensional, and it is composed by the constant functions of $\mathbb{R}$ into $\mathbb{R}^n$. Its image is the space of continuous and T-periodic maps $f : \mathbb{R} \to \mathbb{R}^n$ such that $\int_0^T f(t)\,dt = 0$. Thus it has codimension n in C^0_T. On the other hand, if condition (8.13) is not satisfied, then $G_{(\lambda,u)}$ has infinite-dimensional kernel.

Now, suppose that the Jacobian matrix $\partial_4\phi(t, a, b, c)$ has no negative eigenvalues for every $t \in \mathbb{R}$ and every $a, b, c \in \mathbb{R}^n$. In this case, (8.13) holds for any pair (λ, u). Consequently, $(\widetilde{\Phi}_\lambda)'(u)$ is a Φ_0-operator since it is the sum of the compact linear operator

$$q \mapsto \lambda\partial_2\phi(t, u(t), u(t-\tau), u'(t))q(t) + \lambda\partial_3\phi(t, u(t), u(t-\tau), u'(t))q(t-\tau), \tag{8.14}$$

which is compact since so is the inclusion $C^1_T \hookrightarrow C^0_T$, with the Φ_0-operator $G_{(\lambda,u)}$.

In order to apply the results of the above section, in particular the degree for α-Fredholm maps, we need conditions under which the inequality

$$\alpha(\widetilde{\Psi}, (\lambda, u)) < \omega(\widetilde{\Phi}, (\lambda, u))$$

is satisfied for any given pair (λ, u) in $[0, +\infty) \times C^1_T$ or, at least, in a subset $[0, +\infty) \times \Omega$. The following two results consider some particular and favorable cases in the computation. The first one deals with a modified and more general version Lemma 6.1 in [3].

Proposition 8.6 *Suppose that ψ is Lipschitz-type continuous with respect to the fourth variable, in the following sense: There exist $r > 0$ and $d > 0$ such that*

$$|\psi(t, a, b, c_1) - \psi(t, a, b, c_2)| \leq d|c_1 - c_2| \tag{8.15}$$

for every $t \in \mathbb{R}$, every $a, b \in \mathbb{R}^n$ with $|a|, |b|, |c_1|, |c_2| < r$. Then, denoting by B_r the open ball in C^1_T with center zero and radius r, we have

$$\alpha(\widetilde{\Psi}_\lambda|_{B_r}) \leq \lambda d, \qquad \forall \lambda \in [0, +\infty). \tag{8.16}$$

Proof Fix $\lambda \in [0, +\infty)$. By Corollary 8.1 we have $\alpha(\widetilde{\Psi}_\lambda|_{B_r}) = \lambda\alpha(\Psi|_{B_r})$. Consider now the map

$$V : B_r \times B_r \to C^0_T, \qquad V(u, v)(t) = \psi(t, u(t), u(t-\tau), v'(t)).$$

By the assumptions, we have that $V(u, \cdot)$ is Lipschitz with constant d. Indeed, given $v_1, v_2 \in B_r$ and $t \in \mathbb{R}$, we have

$$|V(u, v_1)(t) - V(u, v_2)(t)| =$$
$$|\psi(t, u(t), u(t-\tau), v_1'(t)) - \psi(t, u(t), u(t-\tau), v_2'(t))| \leq$$
$$d(|v_1'(t) - v_2'(t)|).$$

Hence

$$\|V(u, v_1) - V(u, v_2)\|_\infty \leq d\|v_1 - v_2\|_1.$$

Now, as the extension of V to $C_T^1 \times C_T^1$ is continuous, and due to the compactness of the inclusion $C_T^1 \hookrightarrow C_T^0$, it follows that $V(\cdot, v)$ is completely continuous for any $v \in C_T^1$. Therefore we can apply a slightly extended version of Proposition 13.2 in [12] (see also [16, Exercise 2 in Section 9]), concluding that $\alpha(\Psi|_{B_r}) < d$. Then, the inequality (8.16) follows from Corollary 8.1 as said before. □

Lemma 8.2 *Assume that for any $t \in \mathbb{R}^n$ and any $a, b, c \in \mathbb{R}^n$ the Jacobian matrix $\partial_4\phi(t, a, b, c)$ has no negative eigenvalues. Set*

$$\gamma(\lambda) = \sup_{(t,a,b,c)} \|(I + \lambda\partial_4\phi(t, a, b, c))^{-1}\|. \tag{8.17}$$

Then

$$\omega(\widetilde{\Phi}, (\lambda, u)) \geq \frac{1}{\gamma(\lambda)}, \qquad \forall(\lambda, u) \in [0, +\infty) \times C_T^1.$$

Proof Let $(\lambda, u) \in [0, +\infty) \times C_T^1$ be given. Proposition 8.4 and the fact that $\widetilde{\Phi}$ is C^1 imply $\omega(\widetilde{\Phi}, (\lambda, u)) = \omega\big(\widetilde{\Phi}'(\lambda, u)\big)$. By the same proposition we have $\omega\big(\widetilde{\Phi}'(\lambda, u)\big) = \omega\big((\widetilde{\Phi}_\lambda)'(u)\big)$ and thus

$$\omega(\widetilde{\Phi}, (\lambda, u)) = \omega\big((\widetilde{\Phi}_\lambda)'(u)\big).$$

The assumption on the Jacobian matrix $\partial_4\phi(t, a, b, c)$ implies that condition (8.13) holds for every $(\lambda, u) \in [0, +\infty) \times C_T^1$. Consequently, $(\widetilde{\Phi}_\lambda)'(u)$ is a Fredholm operator of index zero. Since the linear operator defined by the (8.14) is compact, by Proposition 8.3 we have $\omega\big((\widetilde{\Phi}_\lambda)'(u)\big) = \omega\big(G_{(\lambda,u)}\big)$, where $G_{(\lambda,u)}$ has been defined in (8.12). Being, by condition (8.13), $G_{(\lambda,u)}$ invertible, we get, by (5) in Proposition 8.2,

$$\omega\left(G_{(\lambda,u)}\right) = \frac{1}{\alpha\left(\left(G_{(\lambda,u)}\right)^{-1}\right)}.$$

In order to estimate $\alpha\left(\left(G_{(\lambda,u)}\right)^{-1}\right)$, fix $y \in C_T^0$ and let $q \in C_T^1$ be such that $q = \left(G_{(\lambda,u)}\right)^{-1} y$. We have $|q'(t)| \le \|(I + \lambda M_u(t))^{-1}\| \; |y(t)|$ for any $t \in \mathbb{R}$, and thus

$$\|q'\|_\infty \le \max_{t\in\mathbb{R}} \|(I + \lambda M_u(t))^{-1}\| \; \|y\|_\infty \le \\ \sup_{(t,a,b,c)} \|(I + \lambda \partial_4 \phi(t,a,b,c))^{-1}\| \; \|y\| = \gamma(\lambda) \; \|y\|.$$

Therefore, $\|q\|_1 \le \gamma(\lambda) \; \|y\|$. It follows

$$\alpha\left(\left(G_{(\lambda,u)}\right)^{-1}\right) \le \gamma(\lambda).$$

Hence,

$$\omega(\widetilde{\Phi}, (\lambda, u)) = \omega\left(G_{(\lambda,u)}\right) \ge \frac{1}{\gamma(\lambda)},$$

and the proof is complete. □

The above two lemmas are summarized in the next proposition.

Proposition 8.7 *Let $\phi, \psi : \mathbb{R} \times \mathbb{R}^n \times \mathbb{R}^n \times \mathbb{R}^n \to \mathbb{R}^n$ be C^1 and continuous, respectively, and both T-periodic with respect to the first variable. Assume that the following conditions hold:*

(i) ψ verifies the assumption of Lemma 8.6 with respect to suitable $d, r > 0$.
(ii) For any $(t,a,b,c) \in \mathbb{R} \times \mathbb{R}^n \times \mathbb{R}^n \times \mathbb{R}^n$, the Jacobian matrix $\partial_4 \phi(t,a,b,c)$ has no negative eigenvalues.
(iii) d is such that

$$\lambda d < \frac{1}{\gamma(\lambda)} \quad \textit{for any } \lambda \in [0, +\infty), \tag{8.18}$$

where $\gamma(\lambda) = \sup_{(t,a,b,c)} \|(I + \lambda \partial_4 \phi(t,a,b,c))^{-1}\|$.

Then the map

$$H : [0, +\infty) \times C_T^1 \to C_T^0, \qquad H(\lambda, u) = \widetilde{\Phi}(\lambda, u) - \widetilde{\Psi}(\lambda, u), \tag{8.19}$$

where $\widetilde{\Phi}$ and $\widetilde{\Psi}$ are as in (8.11), is an α-Fredholm homotopy in a suitable product $[0, +\infty) \times B_r$, where B_r is the open ball in C_T^1, centered at the origin, with radius r, where r is as in the above condition (i).

As an example illustrating condition (iii) in Proposition 8.7, consider the case in which for any (t,a,b,c) the Jacobian matrix $\partial_4 \phi(t,a,b,c)$ appears in a diagonal form. Suppose that all the eigenvalues are positive and call δ the smallest one. Thus, one can easily check that $\gamma(\lambda) = 1/(1 + \lambda\delta)$, and condition (iii) is clearly satisfied if the Lipschitz constant d of the map ψ is smaller than δ.

Let us come back to our study of problem (8.1). Define

$$L : C^1_T \to C^0_T, \qquad Lu = u'.$$

Problem (8.1) is equivalent to the semilinear operator equation

$$H(\lambda, u) = Lu + \lambda(\Phi(u) + \Psi(u)) = 0 \tag{8.20}$$

in $[0, +\infty) \times C^1_T$, where $H : [0, +\infty) \times C^1_T \to C^0_T$ has also been defined in (8.19). If conditions (i)–(iii) in Proposition 8.7 hold and hence H is an α-Fredholm homotopy, then we can apply the results of Sect. 8.5 to Eq. (8.20) obtaining a global bifurcation result (see Theorem 8.4).

Before stating Theorem 8.4 we need some preliminary remarks. It is not difficult to show that L is a bounded Fredholm operator of index zero. Its kernel is composed by the constant maps that can be (and will be) naturally identified with the elements of $\mathbb{R}^n$. The image of L is

$$\operatorname{Im} L = \left\{ y \in C^0_T : \int_0^T y(t)\, dt = 0 \right\}.$$

Thus, denoting by a slight abuse of the notation, by p either a point in $\mathbb{R}^n$ or a constant map in C^1_T or C^0_T, consider, in the notation of Sect. 8.5, the decomposition $C^0_T = \operatorname{Im} L \oplus \mathbb{R}^n$, and denote by π the associated projection of C^0_T onto $\mathbb{R}^n$. Define the vector field $v : \mathbb{R}^n \to \mathbb{R}^n$ by $v(p) = \pi\Phi(p) + \pi\Psi(p)$. It can be actually written as

$$v(p) = \int_0^T \big(\phi(t, p, p, 0) + \psi(t, p, p, 0)\big)\, dt.$$

We are now ready to state our main result. The statement involves, instead of v, the *mean value vector field* $w : \mathbb{R}^n \to \mathbb{R}^n$ defined by

$$w(p) = \frac{1}{T} \int_0^T \big(\phi(t, p, p, 0) + \psi(t, p, p, 0)\big)\, dt. \tag{8.21}$$

Theorem 8.4 *Let $\phi, \psi : \mathbb{R} \times \mathbb{R}^n \times \mathbb{R}^n \times \mathbb{R}^n \to \mathbb{R}^n$ be C^1 and continuous, respectively. Assume that both are T-periodic with respect to the first variable. Suppose that conditions* (i)–(iii) *in Proposition 8.7 hold. Let $w : \mathbb{R}^n \to \mathbb{R}^n$ be the mean value vector field defined in* (8.21). *Let W be an open subset of $[0, +\infty) \times C^1_T$ and denote $\widetilde{W}_0 = \{p \in \mathbb{R}^n : (0, p) \in W\}$. Assume that the Brouwer degree $\deg_B(w, \widetilde{W}_0, 0)$ is defined and different from zero. Assume, in addition, that H is an α-Fredholm homotopy in W. Then, W contains a connected set of nontrivial solutions of problem* (8.1), *whose closure in W is not compact and intersects $\{0\} \times \operatorname{Ker} L \cong \mathbb{R}^n$ in the compact set $w^{-1}(0) \cap \widetilde{W}_0$.*

Proof The degree $\deg_B(w, \widetilde{W}_0, 0)$ is well defined and not zero if and only if the same is true for $\deg_B(v, \widetilde{W}_0, 0)$. We can apply Theorem 8.3 since $\widetilde{\Phi}$ is orientable. This is a consequence of the fact, proved in [5], that any Fredholm map defined in a simply connected open set (the whole space in this case) is orientable. Thus, the assertion follows from Theorem 8.3. □

Remark 8.2 Clearly, many cases are not so favorable, and the assumptions of Lemmas 8.6 and 8.2 do not hold. Nevertheless, Theorem 8.4 can be applied, for example, in those cases when there exists an open subset U of C^1_T such that H is an α-Fredholm homotopy in $[0, +\infty) \times U$, that is,

$$\alpha(\Psi_\lambda, u) < \omega(\Phi_\lambda, v).$$

If U is simply connected, H is of course orientable in $[0, +\infty) \times U$.

8.7 Examples

In this final section we present two examples illustrating the results of the previous section.

Example 8.5 The first case deals with a parameterized nonautonomous version of the so-called *Mackey–Glass equation*. There is an important literature concerning this kind of equation and its extremely various applications including continuum mechanics. The reader can see, for example, [10, 11, 23]. Here, we consider the equation

$$u'(t) + \lambda \left(\beta(t) \frac{1 + u^2(t-\tau)}{1 + (u^2(t-\tau))^s} u'(t) + \eta(t) f(u(t) + |u'(t)|) \right) = 0,$$

in which $\beta, \eta : \mathbb{R} \to \mathbb{R}$ are C^1 and continuous, respectively, both T-periodic, for a positive given T, while τ is a positive given constant time lag, and $s \in \mathbb{R}$ is greater than 1. The map $f : \mathbb{R} \to \mathbb{R}$ is Lipschitz continuous with constant L. One can observe that, generically, no constant function can solve the equation if $\lambda > 0$. In addition, the equation has not the trivial solution for any positive λ if $f(0) \neq 0$. In the notation of Sect. 8.6, we have $\phi, \psi : \mathbb{R}^4 \to \mathbb{R}$ defined by

$$\phi(t, a, b, c) = \beta(t) \frac{1 + b^2}{1 + (b^2)^s} c \quad \text{and} \quad \psi(t, a, b, c) = \eta(t) f(a + |c|).$$

They are C^1 and continuous, respectively, due to the assumptions on β, η, and f. The definition, in the notation of the above section, of the corresponding Φ and Ψ is straightforward. Let us check if conditions of Proposition 8.7 are satisfied. First of all, we have trivially

$$\partial_4\phi(t, a, b, c) = \beta(t)\,\frac{1 + b^2}{1 + (b^2)^s}.$$

Therefore, for any nonnegative λ, if we assume that β is nonnegative, then any $\widetilde{\Phi}_\lambda$ is a nonlinear Fredholm map of index zero. It is defined in the all space C^1_T, which is simply connected. Then, it is orientable. In addition, a not difficult computation shows that $\gamma(\lambda) = 1$, where $\gamma(\lambda)$ is defined according to (8.17). Concerning the map ψ, recalling Lemma 8.6, it is easy to see that

$$\alpha(\widetilde{\Psi}, (\lambda, u)) \le \lambda L \max_{t\in[0,T]} |\eta(t)|, \qquad \forall(\lambda, u) \in [0, +\infty) \times C^1_T.$$

The combination of these two latest results does not allow us to apply Proposition 8.7, since (8.18) fails. However, if we limit the computation of γ, for a given λ, to b in any interval $(-r, r)$, we see that

$$\frac{1}{\gamma(\lambda)} = \begin{cases} 1 + \lambda \min_{t\in[0,T]} \beta(t) & \text{if } 0 \le r \le 1 \\ 1 + \lambda \min_{t\in[0,T]} \beta(t)\, r & \text{if } r > 1. \end{cases}$$

If we assume that β and η are such that there exists $s > 0$ satisfying

$$\lambda L \max_{t\in[0,T]} |\eta(t)| < \frac{1}{\gamma(\lambda)} \qquad \text{for any } \lambda \in [0, +\infty),$$

in the case when $\gamma(\lambda)$ is computed in the bounded interval $(-s, s)$, then we are in the position to apply Proposition 8.7 in the subset $[0, +\infty) \times B_s$ of $[0, +\infty) \times C^1_T$, where B_s is the open ball with center zero and radius s (recall Remark 8.2).

Now, the mean value vector field defined by formula (8.21) is

$$w(p) = \frac{1}{T}\int_0^T \eta(t)\, dt \cdot f(p).$$

Assuming that the above integral is nonzero, if the Brouwer degree

$$\deg_B(w, I, 0),$$

in open interval $I = (-s, s)$ of $\mathbb{R}$, is well defined and nonzero, we are in the position to apply Theorem 8.4, obtaining the existence of a global bifurcating branch emanating from the set of trivial solutions. The computation of the degree (and its level of difficulty) depends only on the behavior of f in I. □

Example 8.6 The following equation

$$u'(t) = -u(t - \tau) - u^3(t), \tag{8.22}$$

also in perturbed cases, is a delayed version of the so-called *Duffing equation*. Problems of the (8.22)-type are the object of a very reach research in the last decades. We suggest for example the interesting survey of R. Rand in [27, Chapter 3] in which the author also discusses some important applications, including continuum mechanics, of the (8.22)-type equations. In this example, we consider the following parameterized perturbation of (8.22):

$$\begin{cases} u_1'(t) + \lambda \left(u_1(t-\tau) f_1(u_2(t-\tau)u_2'(t)) + h_1(u_1'(t)) + g_1(u_2'(t))u_1^3(t)\right) = 0 \\ u_2'(t) - \lambda \left(u_2(t-\tau) f_2(u_1(t-\tau)u_1'(t)) + h_2(u_2'(t)) + g_2(u_1'(t))u_2^3(t)\right) = 0, \end{cases}$$

in which $f_1, f_2, h_1, h_2 : \mathbb{R} \to \mathbb{R}$ are C^1 and $g_1, g_2 : \mathbb{R} \to \mathbb{R}$ are continuous. Denoting $a = (a_1, a_2)$, $b = (b_1, b_2)$ and $c = (c_1, c_2)$, we have

$$\phi(t, a, b, c) = (b_1 f_1(b_2 c_2) + h_1(c_1), -b_2 f_2(b_1 c_1) + h_2(c_2)) \text{ and}$$
$$\psi(t, a, b, c) = (g_1(c_2)a_1^3, g_2(c_1)a_2^3).$$

Here, C_T^1 stands for the space of the C^1 and T-periodic functions of $\mathbb{R}$ into $\mathbb{R}^2$. For a given $(\lambda, u_1, u_2) \in [0, +\infty) \times C_T^1$, the 2×2 matrix associated to the linear operator $G_{(\lambda, u_1, u_2)}$ defined by the (8.12) is

$$\begin{pmatrix} 1 + \lambda h_1'(u_1'(t)) & \lambda u_1(t-\tau) f_1'(u_2(t-\tau)u_2'(t))u_2(t-\tau) \\ -\lambda u_2(t-\tau) f_2'(u_1(t-\tau)u_1'(t))u_1(t-\tau) & 1 + \lambda h_2'(u_2'(t)) \end{pmatrix}.$$

If we suppose that h_1, h_2 are nondecreasing, a sufficient condition ensuring that this matrix is invertible is $f_1'(t) f_2'(t) \geq 0$ for any real t. With this condition, the nonlinear map $\widetilde{\Phi}_\lambda$ (defined as in the previous cases) turns out to be Fredholm of index zero for any $\lambda \in [0, +\infty)$.

Let us now check the relation between the functions α and ω. The map ψ, because of the growth of the coefficients a_i^3, $i = 1, 2$, only verifies the Lipschitz-type condition (8.15) for a in bounded subsets of $\mathbb{R}^2$, and if g_1 and g_2 are Lipschitz. Assume more precisely that g_1 and g_2 are Lipschitz with positive constant d. We have

$$|\psi(t, a, b, c) - \psi(t, a', b', c')| \leq r^3 d |c - c'|$$

for any given $(t, a, b, c), (t, a', b', c') \in \mathbb{R}^7$, with $a^2 + (a')^2 < r^2$. Therefore, we can say that

$$\alpha(\widetilde{\Psi}, (\lambda, u_1, u_2)) \leq \lambda r^3 d$$

for any $\lambda \in [0, +\infty)$ and any $(u_1, u_2) \in C_T^1$ such that $\|(u_1, u_2)\| < r$.

Let us investigate ω. The study of the matrix

$$(I + \lambda \partial_4 \phi(t, a, b, c)) = \begin{pmatrix} 1 + \lambda h_1'(c_1) & \lambda b_1 b_2 f_1'(b_2 c_2) \\ -\lambda b_1 b_2 f_2'(b_1 c_1) & 1 + \lambda h_2'(c_2) \end{pmatrix}$$

is not particularly simple if we have the purpose to determine $\gamma(\lambda)$ or $\omega(\widetilde{\Phi}, (\lambda, u_1, u_2))$ for a given (λ, u_1, u_2). We can however give some estimates of $\omega(\widetilde{\Phi}, (\lambda, u_1, u_2))$, keeping the assumption that h_1, h_2 are nondecreasing. Recalling that

$$\omega(\widetilde{\Phi}, (\lambda, u_1, u_2)) = \omega(G_{(\lambda, u_1, u_2)}) = \inf_{\|(q_1, q_2)\|=1} \|G_{(\lambda, u_1, u_2)}(q_1, q_2)\|,$$

the following not difficult facts are in order:

1. $\omega(\widetilde{\Phi}, (\lambda, u_1, u_2)) \geq 1$, for every $(\lambda, u_1, u_2) \in [0, +\infty) \times C_T^1$.
2. If we assume that there exists $m > 0$ such that $h_i'(t) \geq m$, $i = 1, 2$, for every $t \in \mathbb{R}$, then $\omega(\widetilde{\Phi}, (\lambda, u_1, u_2)) \geq 1 + \lambda m$, for every $(\lambda, u_1, u_2) \in [0, +\infty) \times C_T^1$.

   Therefore, taking $r = \sqrt[3]{m/d}$, we have

$$\alpha(\widetilde{\Psi}, (\lambda, u_1, u_2)) < \omega(\widetilde{\Phi}, (\lambda, u_1, u_2)) \tag{8.23}$$

for any $\lambda \in [0, +\infty)$ and any $(u_1, u_2) \in C_T^1$ such that $\|(u_1, u_2)\| < r$.

Finally, let as explore the computation of the degree. The vector field $w : \mathbb{R}^2 \to \mathbb{R}^2$ is defined as

$$w(p) = w(p_1, p_2) = (g_1(0)p_1^3 + f_1(0)p_1 + h_1(0), g_2(0)p_2^3 - f_2(0)p_2 + h_2(0)).$$

Consider the first component w_1 of w. If $g_1(0)$ and $f_1(0)$ are not both zero, then the first component of w_1 is proper and its degree $\deg_B(w_1, \mathbb{R}, 0)$ is well defined and is equal to ± 1. The sign coincides with the sign of $g_1(0)$ or, in the case when $g_1(0) = 0$, with that of $f_1(0)$. If the analogous condition holds for the second component of w, then the degree $\deg_B(w, \mathbb{R}^2, 0)$ is well defined and, since w is a diagonal form, one has

$$\deg_B(w, \mathbb{R}^2, 0) = \deg_B(w_1, \mathbb{R}, 0) \cdot \deg_B(w_2, \mathbb{R}, 0) = \pm 1,$$

where the first equality follows immediately from the product formula of the Brouwer degree, since w_1 depends only on p_1 and the same for w_2. The degree of w with respect to a subset U of $\mathbb{R}^2$ needs a more careful analysis. In that case, the degree will depend on the zeros of w in U and their transversality intersection with the real axis. For example, in the notation of Theorem 8.4, a global bifurcating branch is obtained if W is contained in $[0, +\infty) \times B_r$ where the radius r of the ball is as in (8.23), $\widetilde{W}_0$ is a nonempty open subset of the origin in $\mathbb{R}^2$, and the degree $\deg_B(w, \widetilde{W}_0, 0)$ is nonzero. □

References

1. P. Benevieri, A. Calamai, M. Furi, A degree theory for a class of perturbed Fredholm maps. Fixed Point Theory Appl. **2**, 185–206 (2005)
2. P. Benevieri, A. Calamai, M. Furi, A degree theory for a class of perturbed Fredholm maps II. Fixed Point Theory Appl. **2006**, Art. ID 27154, 20 pp (2006)
3. P. Benevieri, A. Calamai, Bifurcation results for a class of perturbed Fredholm maps. Fixed Point Theory Appl. **2008**, Article ID 752657, 19 p (2008)
4. P. Benevieri, M. Furi, A simple notion of orientability for Fredholm maps of index zero between Banach manifolds and degree theory. Ann. Sci. Math. Québec **22**, 131–148 (1998)
5. P. Benevieri, M. Furi, On the concept of orientability for Fredholm maps between real Banach manifolds. Topol. Methods Nonlinear Anal. **16**, 279–306 (2000)
6. P. Benevieri, M. Furi, On the uniqueness of the degree for nonlinear Fredholm maps of index zero between Banach manifolds. Commun. Appl. Anal. **15**, 203–216 (2011)
7. P. Benevieri, A. Calamai, M. Furi, On the degree for oriented quasi-Fredholm maps: its uniqueness and its effective extension of the Leray-Schauder degree. Topol. Methods Nonlinear Anal. **46**(1), 401–430 (2015)
8. P. Benevieri, M. Furi, M. Martelli, M.P. Pera, Atypical bifurcation without compactness. Z. Anal. Anwendungen **24**(1), 137–147 (2005)
9. P. Benevieri, J.G. Mesquita, A. Pereira, Global bifurcation results for nonlinear dynamic equations on time scales. J. Differ. Equ. **269**(12), 11252–11278 (2020)
10. L. Berezansky, E. Braverman, Mackey-Glass equation with variable coefficients. Comput. Math. Appl. **51**(1), 1–16 (2006)
11. L. Berezansky, E. Braverman, On stability of delay equations with positive and negative coefficients with applications. Z. Anal. Anwend. **38**(2), 157–189 (2019)
12. F. Browder, Nonlinear operators and nonlinear equations of evolution in Banach spaces, in *Proceedings of Symposia in Pure Mathematics*, vol. 18/2 (AMS, New York, 1976)
13. A. Calamai, The Invariance of Domain Theorem for compact perturbations of nonlinear Fredholm maps of index zero. Nonlinear Funct. Anal. Appl. **9**, 185–194 (2004)
14. A. Calamai, *A Degree Theory for a Class of Noncompact Perturbations of Fredholm Maps.* PhD Thesis (Università di Firenze, Florence, 2005)
15. G. Darbo, Punti uniti in trasformazioni a codominio non compatto. Rend. Sem. Mat. Univ. Padova **24**, 84–92 (1955)
16. K. Deimling, *Nonlinear Functional Analysis* (Springer, Berlin, 1985)
17. J. Dugundji, An extension of Tietze's theorem. Pacific J. Math. **1**, 353–367 (1951)
18. M. Furi, M. Martelli, A. Vignoli, Contributions to the spectral theory for nonlinear operators in Banach spaces. Ann. Mat. Pura Appl. **118**, 229–294 (1978)
19. M. Furi, M.P. Pera, A continuation principle for periodic solutions of forced motion equations on manifolds and applications to bifurcation theory. Pacific J. Math. **160**, 219–244 (1993)
20. T. Kato, Perturbation theory for linear operators, in *Grundlehren der Mathematischen Wissenschaften*, vol. 132 (Springer, Berlin, 1980)
21. M.A. Krasnoselskii, *Topological Methods in the Theory of Nonlinear Integral Equations* (Pergamos Press, Oxford, 1964)
22. C. Kuratowski, Topologie, in *Monografie Matematyczne*, vol. 20. Warszawa (1958)
23. J. Mallet-Paret, R.D. Nussbaum, A differential-delay equation arising in optics and physiology. SIAM J. Math. Anal. **20**(2), 249–292 (1989)

24. R.D. Nussbaum, Degree theory for local condensing maps. J. Math. Anal. Appl. **37**, 741–766 (1972)
25. G. Prodi, A. Ambrosetti, Analisi non lineare. Quad. **1**. Scuola Normale Superiore di Pisa (1973)
26. P.H. Rabinowitz, Some global results for nonlinear eigevalue problems. J. Func. Anal. **7**, 487–513 (1971)
27. R. Rand, Differential-delay equations, in *Complex Systems, Nonlinear Physics Science* (Higher Ed. Press, Beijing, 2011), pp. 83–117
28. M. Väth, Topological analysis. From the basics to the triple degree for nonlinear Fredholm inclusions, in *De Gruyter Series in Nonlinear Analysis and Applications*, vol. 16 (Walter de Gruyter & Co., Berlin, 2012)

Chapter 9
New Elements for a Theory of Chaos Topology

Denisse Sciamarella and Gisela D. Charó

9.1 Introduction to Topological Analysis

The study of the topological structure of flows in phase space started with a special focus on a number of paradigmatic chaotic attractors. Published more than 60 years ago, Edward Lorenz's famous article in the Journal of Atmospheric Sciences [1] spurred an enormous amount of work on determinism and unpredictability, even though rigorous proof of the existence of the butterfly-shaped attractor remained elusive for decades [2, 3]. The interest in this type of simplified model is interwoven with the desire to understand fundamental issues in physics, such as the complexity of turbulent motion in fluids. Questions of the same nature arose in other fields of science. Otto Rössler, for instance, tried to reproduce the Lorenz attractor in a chemical reaction, discovering that "chaos" could be produced with a system that was even simpler than the Lorenz attractor [4]. This single-spiral attractor bears Rössler's name and is used in many studies, e.g., [5], as a prototype of the simplest possible chaotic behavior, just as the hydrogen atom is the simplest element in Mendeleev's table. A computer-assisted proof of its existence dates back to only a couple of years ago [6]. Research, not only on chaotic attractors, but on the structure underlying a flow in phase space in general, proliferated independently

D. Sciamarella (✉)
Institut Franco-Argentin d'Études sur le Climat et ses Impacts (IFAECI - CNRS, CONICET, IRD, UBA), Buenos Aires, Argentina
e-mail: denisse.sciamarella@cnrs.fr

G. D. Charó
Centro de Investigaciones del Mar y de la Atmósfera (CIMA-FCEN, UBA), Buenos Aires, Argentina
e-mail: gisela.charo@cima.fcen.uba.ar

P. Amster, P. Benevieri (eds.), *Topological Methods for Delay and Ordinary Differential Equations*, Advances in Mechanics and Mathematics 51,
https://doi.org/10.1007/978-3-031-61337-1_9

of the existence of rigorous proofs, because of its relevance to understand the mechanisms responsible for complexity in nonlinear deterministic systems.

In 1983, Rössler rose the question of whether it was possible to construct a taxonomy of chaotic attractors, i.e., a general scheme with classes of qualitatively different chaotic behavior in phase space [7]. The challenge led to the study of the structure underlying the numerical solutions to different equation systems and of its properties, whether geometrical or topological. The very first attempts to describe the topology of a flow structure were limited to the use of knot theory which is inherently three dimensional and only later extended to higher-dimensional phase spaces. An up-to-date review on dynamical systems and algebraic topology, with special attention to the applications in the climate sciences, can be found in [8].

Homology groups were the natural candidate that held promise during decades to aid in the construction of a topological taxonomy in an arbitrary number of phase space dimensions. Cell complexes were shown to work successfully as a topological descriptor of a numerical solution to a set of four-dimensional ODEs [9]. As mentioned therein,

> the correspondence between a numerical solution and a cell complex with certain topological properties cannot be taken as an indication of chaoticity. Yet, it provides a global description of the way in which the phase space should be mapped into itself under the action of the correct model equations.

Such a correspondence holds because a cell complex approximating a numerical solution (or a flow reconstructed from data) captures, to some extent, the stretching and folding processes that shape the solutions of a deterministic set of equations in phase space. Cell complexes were successfully used to identify, for instance, classes of finite-time particle behavior in an incompressible fluid [10, 11]. The approach enables detecting subsets of fluid particles moving together without mixing with the surroundings. Tagging the behavior of particles enables distinguishing the nonmixing regions in a fluid. This is true independently of whether such a behavior is or is not chaotic. In all cases, the description is done through homology group and torsion group generators. But this approach only provides a partial solution to the taxonomy problem, since it does not include the underlying flow in the description.

This chapter is devoted to the development of a new mathematical object capable of describing both, the structure in phase space and the nonequivalent ways to flow along such structure. This object is now called a templex [12] and can be seen as a kind of dynamic *kirigami*, a cut-out drawing made of pieces that fit together, taking into account the direction of flow.

9.2 Flows in Phase Space

In dynamical systems theory, given a differential equation of the form:

$$\frac{d\boldsymbol{x}}{dt} = \dot{\boldsymbol{x}} = \boldsymbol{f}(\boldsymbol{x}, t), \tag{9.1}$$

where time is the only independent variable, $\boldsymbol{x}(t) \in \mathbb{R}^n$, and $\boldsymbol{f} : \mathbb{R}^{n+1} \to \mathbb{R}^n$ the smooth differential map that generates a *flow* or *phase flow*, i.e., a diffeomorphism $\boldsymbol{\phi}(t, \boldsymbol{x})$ that satisfies Eq. (9.1). Phase space is the space spanned by the dependent variables $\boldsymbol{x}$.

One of the best known methods of analyzing a phase flow is metric in nature: The quantities describing the fractal properties of the invariant set to which the observed trajectory belongs (e.g., the fractal dimension) allow estimating, for instance, the degree of complexity of the flow. However, this quantification of complex solutions does not allow to perform a proper classification of nonlinear dynamical systems in terms of well-defined classes. For example, the Rössler system

$$\begin{aligned}\dot{x} &= -y - z\\ \dot{y} &= ay + x\\ \dot{z} &= b + z(x - c),\end{aligned}$$

with $a = 0.43295, b = 2, c = 4$ and the Lorenz '63 system:

$$\begin{aligned}\dot{x} &= \sigma(y - x)\\ \dot{y} &= x(\rho - z) - y\\ \dot{z} &= xy - \beta z,\end{aligned}$$

with $\rho = 28, \sigma = 10, \beta = \frac{8}{3}$. Having essentially different dynamics, these attractors have very close values of fractal dimension: 2.06 ± 0.01 for the Lorenz and 2.01 ± 0.01 for Rössler case. With some noise contamination, these two dynamics would not be easy to distinguish using fractal dimensions. The chaotic attractors are shown in Fig. 9.1. For a historical perspective on these two sets of ODEs, the reader is referred to [13, 14].

In this text, we will be concerned by a topological description of the structure of chaotic flows in phase space, which will enable distinguishing between nonequiv-

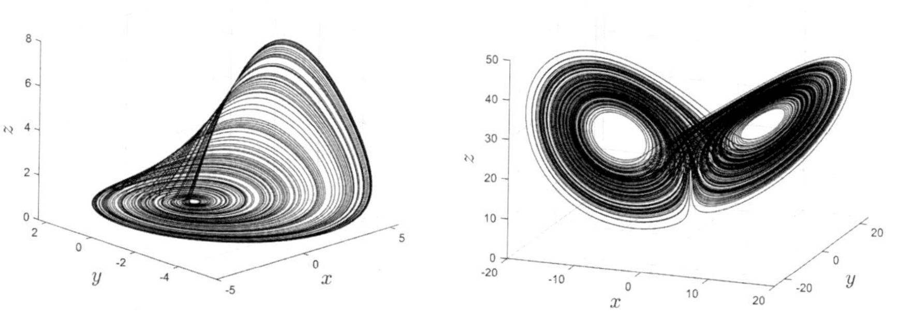

Fig. 9.1 Rössler and the Lorenz chaotic attractors, produced by a folding and a tearing mechanism, respectively, exhibiting qualitatively different dynamics

alent dynamics. Interest in the manner in which the neighborhood of a flow is structured in phase space has led scientists through different roads. Here we will make a tour of necessary definitions leading to what we now call a *templex*, the mathematical object tailored to describe the structure of a phase flow. As we shall see, this description will require a mathematical representation for the structure itself, but also of the different manners in which this structure can be effectively traveled by the flow.

The first attempts to achieve this description were based on observations concerning numerical solutions to nonlinear systems. Aperiodic solutions can sometimes be indistinguishable from long periodic orbits: If the purpose is to describe the organization of trajectories within an attractor in phase space, approximating an aperiodic orbit with a periodic solution may be of help [15]. The closed orbits approximating the unstable periodic orbits (UPOs) can be regarded as the skeleton around which a chaotic flow is organized [16–18]. Converting these orbits into knots and investigating them in terms of knot invariants, such as linking numbers or relative rotation rates, constitute a step further [19]. Such knot invariants are finally synthesized under the form of a knot holder or template [20, 21], which will become a key element in a field that is identified as "chaos topology" or "topology of chaos."

The knot holder is defined as a branched 2-manifold with boundary, embedded in $\mathbb{R}^3$. A branched manifold has the property that each point in it has a neighborhood that is homoeomorphic to either a full ball or a half ball, thus allowing for the presence of boundaries [22]. We do not require a branched manifold to be Hausdorff. A branching line is an example of a non-Hausdorff manifold: It has a fork shape at the juncture. A branched manifold can thus have branches heading toward different parts of phase space. Proving that every finite subset of closed orbits in the Lorenz flow projects onto a specific knot holder required results that would become available more than two decades later [2, 23, 24].

Even if the terms branched manifold, knot holder, and template are often used interchangeably, the concept of *template* as introduced in [25] involves the semiflow, i.e., the oriented flow on the branched surface with boundaries whose orientation is not reversible. The manner in which this semiflow is structured is the result of a sequence of stretching, folding, tearing, and squeezing processes that occur repetitively for any set of initial conditions in phase space under the action of the equation system [26].

A template is the representation of a branched manifold in the form of a standard chart, where the branches, the different paths allowed by the flow, are represented as strips that preserve the crossing and twisting information of the flow. A strip is defined between a splitting chart, where the strips are fictively split (fictive holes), and a joining chart, where the strips are joined into a single one. Figure 9.2 shows the templates for the Rössler and Lorenz systems of Fig. 9.1.

The algebraic description of a template is expressed by the linking matrix $L \in \mathbb{R}^{N_s \times N_s}$, where N_s is the number of strips of the branched manifold, L_{ii} is the sum of the signed half-twist of the strip i, and L_{ij} is the sum of the permutations between strips i and j ($i \neq j$). In the case of the Rössler template (see Fig. 9.2), there are two strips $N_s = 2$. Strip 1 is a simple turn without twist $L_{11} = 0$, while strip 2

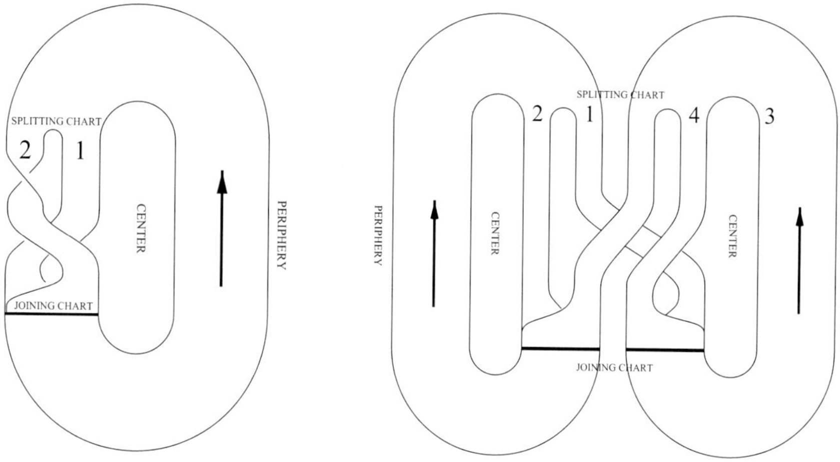

Fig. 9.2 Templates representing the twists and crossings of the two and four strips, respectively, for the Rössler (left) and the Lorenz (right) attractors

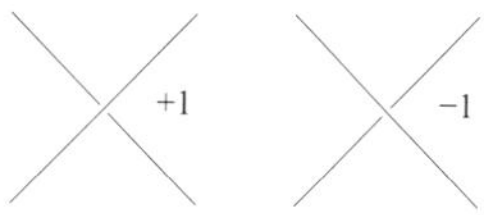

Fig. 9.3 The standard crossing convention for the sign of the half-twist of the strips in a template

has a negative half-twist $L_{22} = -1$ (Fig. 9.3). A negative crossing $L_{12} = -1$ of strips 1 and 2 is imposed by the standard insertion rules [27–29]. According to this definition, the linking matrices for the Rössler (L_1) and Lorenz (L_2) attractors read:

$$L_1 = \begin{bmatrix} 0 & -1 \\ -1 & -1 \end{bmatrix}$$

$$L_2 = \begin{bmatrix} 0 & 0 & 0 & 0 \\ 0 & +1 & 0 & 0 \\ 0 & 0 & 0 & 0 \\ 0 & 0 & 0 & +1 \end{bmatrix}.$$

The relevance of these templates regarding the fundamental question of whether we can classify the various types of behavior of deterministic dynamical systems [30]—as much as Dmitri Mendeleev classified elements according to their chemical properties—is undeniable. Templates have also been applied to several scientific problems [31–37]. The main drawback with templates, however, is that they can only be constructed in three-dimensional phase spaces. Constructing them from a set of trajectories also relies on the possibility of reconstructing good approximations of UPOs coexisting with an attractor. This may become penalizing in the presence of short or noisy time series, as when the trajectories are the result of experimentally measured variables. All these limitations motivated the search for "knotless" or "braidless" methods [38], i.e., of topological invariants that would not become trivial in higher dimensions. The authors in [22] had already made a step forward, using the Euler characteristic and Betti numbers to describe a limit cycle and a torus in phase space. The construction of a cell complex as defined in algebraic topology [39] faithfully representing the flow on a branched manifold in phase space was first attempted in [9, 40]. The next section will describe the particular features of such a cell complex.

9.3 Branched Manifold Analysis Through Homologies (BRAMAH)

A cell complex that correctly approximates the topological structure of the Rössler and Lorenz attractors can be constructed assuming that the point cloud under analysis lies on a branched 2-manifold. The cells of such a complex (a set of polygons) will be such that the subsets of points approximated by the cells are locally homeomorphic to either a full disk or a half disk.

Definition 9.1 Let ϕ be a sufficiently well-sampled phase flow with an underlying d-manifold in $\mathbb{R}^n$. A BRAMAH complex of dimension h has k-cells ($k \leq n$) such that:

(i) The 0-cells are a sparse subset of the original cloud of points in phase space.
(ii) $h = d$ with d the local dimension of the branched manifold ($d \leq n$).
(iii) Each h-cell is such that h of the singular values describing the distribution of the points around the barycenter of the h-cell scale linearly with the number of points in the cell.

The third property holds as long as the effects of local curvature in the branched manifold become apparent: The singular values that scale as higher powers measure the deviation from the tangent space.

Many different BRAMAH complexes can be built from a point cloud depending on the tolerance to curvature that is set in the computation, but all of them will comply with the property of providing a good skeleton for the structure that supports the semiflow. The geometry in the dynamic description is in fact being taken as input for the topological computations. Notice that the 0-cells of the cell complex are associated to a set of points in the original dataset that is used to construct the cell complex. In this respect, the approach leads to a structure that is immersed in full phase space, offering a description of the intrinsic as well as the extrinsic topological properties of the attractor.

Figure 9.4 shows examples of BRAMAH simplified complexes (having a low number of cells) for the Rössler (K_R) and Lorenz (K_L) point clouds considered in each case. Planar diagrams for these complexes are shown in Fig. 9.5, where

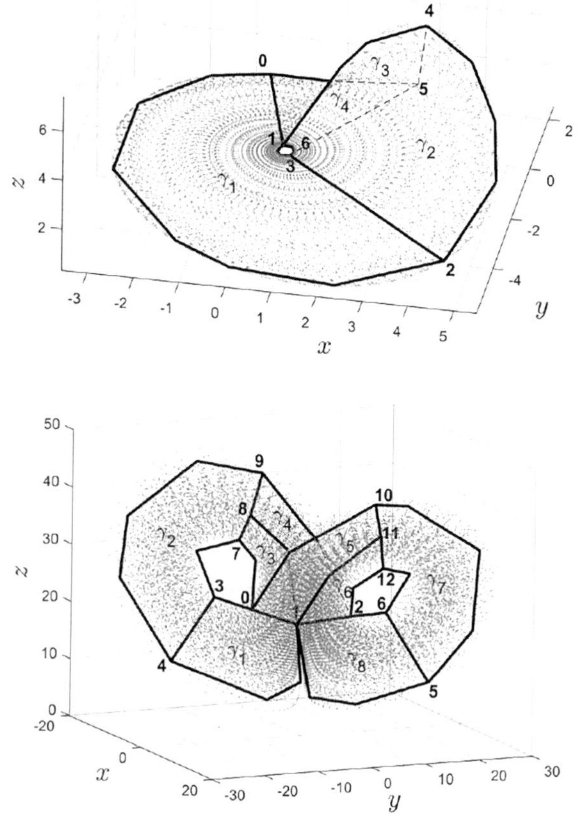

Fig. 9.4 Simplified BRAMAH cell complexes K_R, K_L for the Rössler (top) and the Lorenz (bottom) attractors

duplicated indices denote gluing prescriptions, as detailed in the caption. BRAMAH complexes are uniformly oriented, and one chooses an orientation for one 2-cell (clockwise or anticlockwise) and propagates it across the rest of the 2-cells.

Rössler Branched Manifold

The BRAMAH complex K_R for the Rössler attractor (first image in Figs. 9.4 and 9.5) has the following 2-cells: $\gamma_1 = \langle 0, 2, 3, 1\rangle$, $\gamma_2 = \langle 3, 2, 4, 5, 6\rangle$, $\gamma_3 = \langle 6, 5, 0, 1\rangle$, $\gamma_4 = \langle 5, 4, 1, 0\rangle$. The 2-cells are spelled here in terms of the 0-cells and directed in the sense in which the vertices are listed (anticlockwise). The directions are also shown in the figure with arrows. The 1-cells are also spelled in terms of the 0-cells. The direction of the 1-cell is taken to be positive when the index of the departure 0-cell is lower than the arrival 0-cell. A k-chain C_k is defined as

$$C_k = \sum_i a_i \, \sigma_k^i,$$

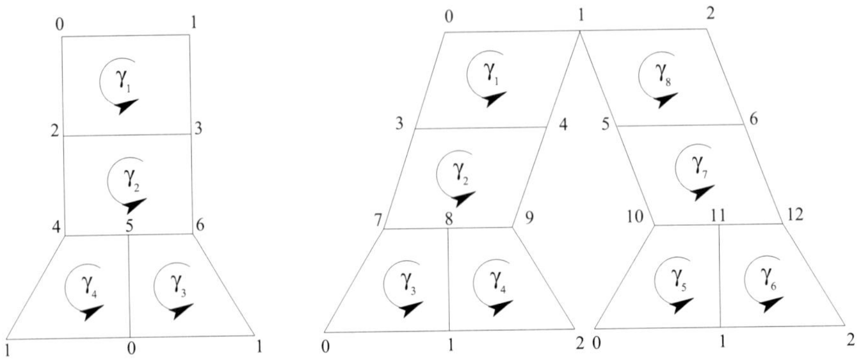

Fig. 9.5 Planar diagrams of complexes K_R and K_L. Repeated indices prescribe gluing. For instance, the 1-cell $\langle 0, 1\rangle$ in K_R appears thrice, in γ_1, γ_3, and γ_4. This means that these three 2-cells are glued by $\langle 0, 1\rangle$

where $a_i \in \mathbb{Z}$ and σ_k^i is a k-cell ($\forall i \in \mathbb{N}$). When we apply the border operator $\partial_k : C_k \to C_{k-1}$, we have: $\partial_1(\langle 0, 1\rangle) = \langle 1\rangle - \langle 0\rangle$ and $\partial_2(\gamma_1) = -\langle 0, 1\rangle - \langle 1, 3\rangle + \langle 2, 3\rangle + \langle 0, 2\rangle$.

Computing the homology groups for K_R is quite simple. A Wolfram Mathematica© code is available in https://git.cima.fcen.uba.ar/sciamarella/templex-properties, to reproduce all the computations discussed in this chapter. We obtain $H_0 \sim \mathbb{Z}^1$, $H_1 \sim \mathbb{Z}^1$, and $H_2 \sim \emptyset$. The generator of H_1, also called a 1-hole, can be written as a chain of 1-cells: $-\langle 0, 1\rangle$+$\langle 0, 2\rangle$+$\langle 2, 4\rangle$-$\langle 1, 4\rangle$. If one locates this generator in the phase space, it is easy to see that it goes around the unstable focus (fixed point) of the ODE system. The second fixed point causes the rise and twist in the z dimension, but this is not captured by homology groups.

Let be Γ the sum of the 2-cells in a complex K, and consider

$$\partial_2(\Gamma) = \sum_j a_j \sigma_1^j + \sum_j b_j \sigma_1^j,$$

where $a_j = \pm 1$, $b_j \neq \pm 1$, and σ_1^j are 1-cells in K. The first term refers to the boundary of the manifold underlying K, and if it is zero, the underlying manifold has no boundary. The second term indicates the torsion elements, and if it is nonzero, the underlying manifold is said to be nonorientable.

The gluing information of cells γ_3 and γ_4 to γ_1 in K_R indicates that this structure is not equivalent to other single 1-holed structures as the cylinder or the Möbius band, in spite of their being homologically equivalent. Of course, there is more information in a cell complex than the one conveyed by its homology groups. Torsion groups, for instance, are also of interest, since they enable distinguishing a cylinder from a Möbius band. These properties can be computed along with homology groups, to better describe the topological structure of a manifold.

Another piece of information that is extremely relevant to a branched 2-manifold is the existence of joining loci.

Definition 9.2 Let K be BRAMAH two-dimensional complex and let γ_i denote its 2-cells. The **joining locus** (or loci) is formed by the 1-cells shared by at least three 2-cells. The 1-cells and 2-cells with this property are called joining 1-cells and joining 2-cells, respectively.

In the Rössler example, $\langle 0, 1\rangle$ is a joining 1-cell, and γ_1, γ_2, and γ_4 are joining 2-cells. These cells carry information regarding the branching of the manifold.

Lorenz Branched Manifold

The Lorenz BRAMAH complex K_L is not homologically equivalent to the Rössler case: $H_0 \sim \mathbb{Z}^1$, $H_1 \sim \mathbb{Z}^2$, and $H_2 \sim \emptyset$. Two 1-holes are present in this structure, one per wing of the butterfly. They are given by $+\langle 1, 5\rangle + \langle 5, 10\rangle - \langle 0, 10\rangle + \langle 0, 1\rangle$ and $+\langle 0, 3\rangle + \langle 3, 7\rangle - \langle 0, 7\rangle$. These 1-holes encircle the unstable foci of the ODE system. The remaining fixed point of the system, of a saddle type, is not detected with homologies. No torsion groups are found for the complex.

The branched nature of this branched 2-manifold is detectable looking for the joining locus, defined by the joining 1-cells $\langle 0, 1\rangle$, $\langle 1, 2\rangle$ and by the joining 2-cells $\{\gamma_1, \gamma_3, \gamma_5\}$ and $\{\gamma_4, \gamma_6, \gamma_8\}$, respectively. Coming back to the phase space, the joining locus is located in the rotation axis, where we can also find the saddle node (the third fixed point) of the ODEs.

! Important

In this description of branched manifolds with BRAMAH cell complexes, an important piece of information is missing to complete the portrait of a dynamical system with the same level of detail offered by templates for three-dimensional flows. Templates encode the knot content of an attractor, but this is done taking the flow into account through the splitting and joining charts, and through the introduction of fictive holes, which cannot necessarily be classified as holes in the sense of homologies. Unlike knots on their own, templates are knot organizers that take the timeline into account. But cell complexes and their properties—homologies, torsions, weak boundaries, and even joining lines—have not been conceived as descriptors of structures in phase space, but as mathematical objects describing a manifold—or a branched manifold—in a space *tout court*.

As discussed in [41], bringing time into the phase space is sometimes done considering an additional axis t: This "space" is sometimes called *extended phase space*. The procedure overlooks, however, a fundamental fact in dynamical system's theory, which assumes that state variables are dependent, while time is the only independent variable of the system [42]. Using an extra axis for time is tantamount to endowing time with a dual status: that of an independent variable and of a dependent variable simultaneously. This is the cause of many severe difficulties

[43], one of which is that the phase space is no longer bounded. Let us recall that boundedness is—together with sensitivity to initial conditions—an essential ingredient to produce chaos.

In this chapter, we will incorporate time into the topological description of an attractor in $\mathbb{R}^n$ with $n \geq 3$, without altering the implicit character of the sole independent variable that distinguishes phase space from any other form of mathematical space. This will be achieved introducing a new object to describe the flow on an attractor in phase space. This object will be called a "templex" and will be defined, described, and applied to a few examples in the sections that follow.

9.4 Templex

When a cell complex is used to describe a structure of a flow in phase space, a basic observation stands out. In a cell complex, a path can be denoted using chains of cells. Any path is possible; some of these paths will be homologically equivalent, others will not, but none of them is forbidden. This is true for a space without time. But in phase space, an arbitrary path will not necessarily conform with the flow.

In K_R, for instance, from the 2-cell γ_1, the flow path can only go toward γ_2, and from γ_2 can go either to γ_3 or to γ_4, but it would be impossible to go from γ_2 toward γ_1. In a template, the flow cannot be reversed at the joining chart without violating uniqueness. The existence of time or, equivalently, of a flow will restrict the possibility of circulating along the cells of a complex.

The directions imposed by the flow must not be confounded with the directions or orientations of the cells. Let us recall that a two-dimensional complex is directed if every 1-cell is given a direction (from initial point to terminal point) and each 2-cell is set to be clockwise or counterclockwise. These definitions are necessary to introduce an algebra of chains, but they tell us nothing about possible rules to travel along the 2-cells of the complex according to the flow. The connections between 2-cells that are allowed by the flow must be informed separately, using, for instance, a directed graph or digraph. For this digraph to encode the possible paths along the 2-cells of a two-dimensional complex, its nodes must be associated to the indices of the 2-cells, and the edges must denote the allowed transitions between them.

Such a digraph can be computed directly from the attractor dataset using not only the coordinates of the points in phase space $(x(t_i), y(t_i), z(t_i))$ but also the sequential information given by time: $(x(t_{i+1}), y(t_{i+1}), z(t_{i+1}))$. Let us construct the digraphs for the complexes K_R and K_L of the last section and plot them in Fig. 9.6. It is easy to see in K_R that γ_1 only leads to γ_2, while γ_2 may lead to γ_3 or γ_4; γ_3 then leads back to γ_1 as well as γ_4. The digraph for K_L can be interpreted in the same manner.

The complex and the digraph form an object that will enable us to extract the properties that are encoded in a template using a cell complex. Such an object will be called templex out of the contraction of the two keywords "template" + "complex." The advantage of a templex (over a template) is methodological—it

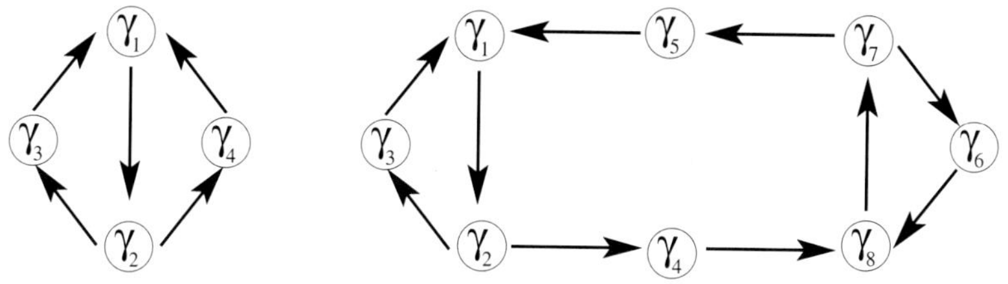

Fig. 9.6 Digraphs D_R and D_L for K_R (left) and K_L (right). The nodes of the digraphs are indexed using the numbers of the γ_i 2-cells in each case

directly describes the branched manifold instead of describing the knotted unstable periodic orbits—and theoretical (since the description is not subjected to the dimensional restrictions that are proper to knots).

Definition 9.3 A **templex** $T \equiv (\overline{K}, D)$ of dimension d is made of a d-complex K approximating a branched d-manifold associated with a flow in phase space, and a digraph $D = (V, E)$ such that (i) the nodes V are the d-cells of $\overline{K}$ and (ii) the edges E are the flow-allowed connections between the d-cells.

In K_R and K_L, we have $d = 2$. The following paragraphs will provide further details on how to extract the properties of $T_R \equiv (\overline{K}_R, D_R)$ and $T_L \equiv (\overline{K}_L, D_L)$. Computing these properties from complexes and the digraphs requires some extra work if the cells of the complex participating in the joining loci are not minimal in number. We will not describe this procedure here since $\overline{K}_R$ and $\overline{K}_L$ already satisfy this condition. For details on this procedure, the reader is referred to [12], where the templex is introduced for the first time.

Definition 9.4 Let $T = (\overline{K}, D)$ be a templex and D_s the subgraph of D, whose nodes are the joining 2-cells of the flow-oriented complex $\overline{K}$. A **joining node** is defined as the node whose number of edges $\subset D_s$ leading into that node (in-degree) is greater than or equal to 2. The joining 2-cells are also the joining nodes, and their respective joining 1-cells are called **templex joining loci**.

In the templex joining loci, we will identify the *ingoing* and *outgoing* 2-cells, i.e., the 2-cells that are visited by the flow just before or just after crossing one of the templex joining 1-cells, respectively. The only outgoing 2-cell in T_R is γ_1, and its in-degree as a node is equal to two (the ingoing 2-cells are γ_4 and γ_3). The outgoing 2-cells in T_L are γ_1 and γ_8, whose in-degree is also equal to two, since γ_1 is fed by γ_3 and γ_5, and γ_8 is fed by γ_4 and γ_6.

To conform the templex to the same flow-related conventions used for templates, we will impose a number of additional conditions on how the cells of the complex are oriented. The flow-oriented versions $\overline{K}_R$ and $\overline{K}_L$ of the complexes in Fig. 9.5 are provided in Fig. 9.7. A cell 2-complex is said to be flow-oriented if the 2-cells are oriented from the center to the periphery [44], where the center of an attractor is associated with the focus point around which trajectories circle. The

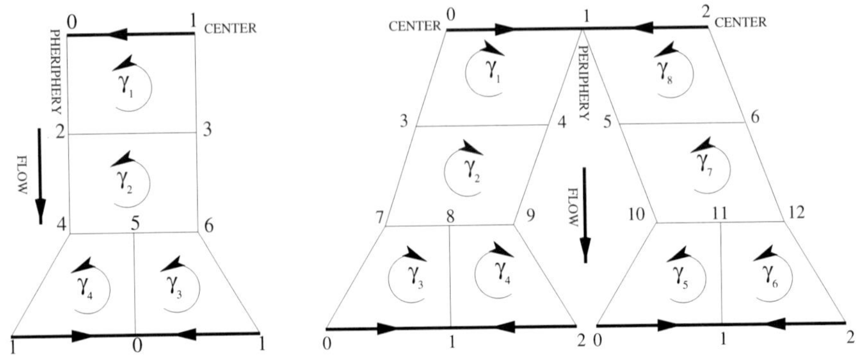

Fig. 9.7 Flow-oriented cell complexes $\overline{K}_R$ and $\overline{K}_L$. The thick lines mark the 1-cells that are now oriented in agreement with the flow-oriented 2-cells. The direction of the flow is indicated with the side arrows

joining 1-cells (thick arrows in Fig. 9.7) inherit the orientation of the outgoing 2-cell from which they are a border. Let us consider $\overline{K}_R$ first: as γ_1 is counterclockwise, the orientation of the joining 1-cell $\langle 0, 1\rangle$ is set to go from $\langle 1\rangle$ to $\langle 0\rangle$. The templex joining locus is said to have one component: $J_1 = -\langle 0, 1\rangle$. In $\overline{K}_L$, the joining 1-cells $\{\langle 0, 1\rangle, \langle 1, 2\rangle\}$ are oriented so that $\langle 0, 1\rangle$ goes from $\langle 0\rangle$ to $\langle 1\rangle$ and $\langle 1, 2\rangle$ goes from $\langle 2\rangle$ to $\langle 1\rangle$. The Lorenz flow is split at the 0-cell connecting the oppositely directed joining 1-cells; $\langle 1\rangle$ is said to be a splitting 0-cell. The templex joining locus has two components: $J_1 = +\langle 0, 1\rangle$ and $J_2 = -\langle 1, 2\rangle$. This result is compatible with the fact that the Poincaré section of the Lorenz attractor has two components [12].

In a template, the strips indicate the number of nonequivalent paths around the branched manifold. In a templex, this information can be extracted looking for cycles in the digraph that start and end in the outgoing 2-cells of the templex joining locus. As the nodes of the digraph are 2-cells of the cell complex, such cycles will immediately denote tracks of 2-cells in the complex $\overline{K}$.

Definition 9.5 A **generatex** $\mathcal{G} = (\overline{K}^g, D^g)$ is a subtemplex in $T = (\overline{K}, D)$, where D^g is a cycle of the digraph D, and $\overline{K}^g \subset \overline{K}$. A generatex is said to be of order p with $p \in \mathbb{N}$, $p \geq 1$, if the cycle has p distinct ingoing nodes. A generatex with $p = 1$ will be said to be *simple*, and *degenerated* otherwise.

The number of generatexes will depend on the number of cells in the complex, but the tracks of cells that start and finish in the same outgoing cell (cycles) will be considered equivalent or redundant. The union of all equivalent generatexes (associated with equivalent cycles) plays a fundamental role in the characterization of branched manifolds. We will thus retain a single representative for each group of equivalent generatexes, and we will refer to them as representative generatexes. In the examples we are using here, the structure of the complexes is simplified, and there are therefore no redundant cycles.

One more step is necessary before we can define the analog of a strip in a templex, since strips are always simple—in the sense that they are defined so that they do not pass through the joining locus more than once. This is so because a strip in a template is related to the structure of the first-return map to the Poincaré section. If a templex is such that all generatexes are simple, we are already done. But if there are generatexes of order $p > 1$, we will need to split them into p pieces to get to the analog of a strip. This will be achieved by introducing the definition of weak cycle. The order p of a cycle in the digraph D^g is given by the number of ingoing nodes that correspond to the ingoing 2-cell of the subcomplex $\overline{K}^g$. A generatex of order p can be considered the union of p weak cycles.

Definition 9.6 Let $\mathcal{G}_i$ be the i-th representative generatex of templex $T = (\overline{K}, D)$. If $\mathcal{G}_i$ is of order one, it is said to be a strong **stripex** $\mathcal{S}_i$. If $\mathcal{G}_i$ is degenerated, with order $p > 1$, $p \in \mathbb{N}$, it is said to be composed by p weak stripexes $\mathcal{S}_{i,j}$ with $1 \leq j \leq p$.

Locating stripexes in the phase space is straightforward for a BraMAH complex, since the coordinates of the 0-cells in the phase space are available by construction. The free edges of a stripex $\mathcal{S}_i$ are composed by the 1-chains of the associated subcomplex $\overline{K}_i$ that result from applying the boundary operator to the sum of all the cells in $\overline{K}_i$ with the exception of those composing the joining locus. The stripex will be said to have a local twist if the free edges of $\overline{K}_i$ (see Fig. 9.8) change their relative positions with respect to the orientation from the center to the periphery. In the analogy with templates, uneven local torsions in a strip correspond to a local twist in a stripex.

A code used to automatically extract the stripexes from the list of 2-cells in a flow-oriented BraMAH complex $\overline{K}$ and the list of nodes and edges in the associated digraph D forming the templex $T = (\overline{K}, D)$ (with a minimal structure in the joining loci) is provided in https://git.cima.fcen.uba.ar/sciamarella/templex-properties.

Rössler Stripexes

The templex $T_R = (\overline{K}_R, D_R)$ for the Rössler attractor has two simple generatexes $\{\mathcal{G}_{R_1}, \mathcal{G}_{R_2}\}$, i.e., two strong stripexes $\{\mathcal{S}_{R_1}, \mathcal{S}_{R_2}\}$. Figure 9.6 shows $\mathcal{S}_{R_1}, \mathcal{S}_{R_2}$ in the cell complex: They correspond to strips 1 and 2 in the template of Fig. 9.2.

$\mathcal{G}_{R_1} = \mathcal{S}_{R_1} = (\overline{K}_{R_1}, D_{R_1})$ $\overline{K}_{R_1}$ formed by the 2-cells $\{\gamma_1, \gamma_2, \gamma_4\}$ and $D_{R_1} = \{1 \to 2,\ 2 \to 4,\ 4 \to 1\}$

$\mathcal{G}_{R_2} = \mathcal{S}_{R_2} = (\overline{K}_{R_2}, D_{R_2})$ $\overline{K}_{R_2}$ formed by cells $\{\gamma_1, \gamma_2, \gamma_3\}$ and $D_{R_2} = \{1 \to 2,\ 2 \to 3,\ 3 \to 1\}$

The first stripex $\mathcal{S}_{R_1}$ has no twists, and the second one $\mathcal{S}_{R_2}$ has a twist (Fig. 9.8). If $\{\overline{K}_{R_1}, \overline{K}_{R_2}\}$ are analyzed as independent complexes, they have the properties of a Möbius band and of a cylinder, respectively. This is why this attractor is often described as resulting from a combination of a Möbius band and an ordinary band (cylinder), glued together side by side [4].

Lorenz Stripexes

The templex $T_L = (\overline{K}_L, D_L)$ for the Lorenz attractor has three generatexes $\{\mathcal{G}_{L_1}, \mathcal{G}_{L_2}, \mathcal{G}_{L_3}\}$ associated with the cycles:

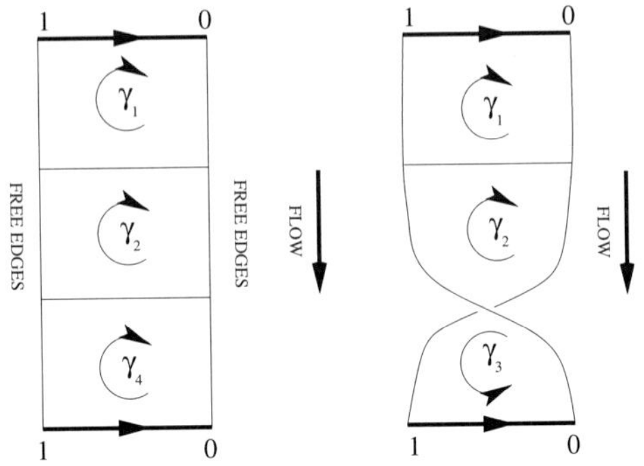

Fig. 9.8 Subcomplexes $\overline{K}_{R_1}$ (left), $\overline{K}_{R_2}$ (right) corresponding to stripexes $\mathcal{S}_{R_1}$ and $\mathcal{S}_{R_2}$, respectively. The free edges are indicated in $\overline{K}_{R_1}$

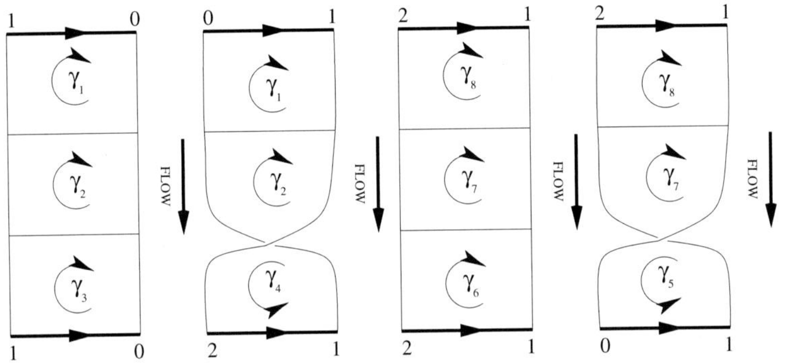

Fig. 9.9 Subcomplexes $\overline{K}_{S_1}$, $\overline{K}_{S_2}$, $\overline{K}_{S_3}$, and $\overline{K}_{S_4}$ (from left to right) corresponding to stripexes $\mathcal{S}_{R_1}$, $\mathcal{S}_{R_2}$, $\mathcal{S}_{R_3}$, and $\mathcal{S}_{R_4}$, respectively

$\mathcal{G}_{L_1}$ $\quad c_1 \equiv 1 \to 2 \to 3 \to 1$
$\mathcal{G}_{L_2}$ $\quad c_2 \equiv 1 \to 2 \to 4 \to 8 \to 7 \to 5 \to 1$
$\mathcal{G}_{L_3}$ $\quad c_3 \equiv 8 \to 7 \to 6 \to 8$

These generatexes yield four stripexes $\{\mathcal{S}_{L_1}, \mathcal{S}_{L_2}, \mathcal{S}_{L_3}, \mathcal{S}_{L_4}\}$, corresponding to strips $1, 2, 3, 4$ in the template of Fig. 9.2. Two of the generatexes $\{\mathcal{G}_{L_1}, \mathcal{G}_{L_3}\}$ are simple and lead to the strong stripexes $\{\mathcal{S}_{L_1}, \mathcal{S}_{L_3}\}$. The generatex $\mathcal{G}_{L_2}$ is degenerated (of order 2) and leads to the weak stripexes $\{\mathcal{S}_{L_2}, \mathcal{S}_{L_4}\}$.

$\mathcal{S}_{L_1} = (\overline{K}_{L_1}, D_{L_1})$ $\quad \overline{K}_{L_1}$ formed by cells $\{\gamma_1, \gamma_2, \gamma_3\}$ and $D_{L_1} = \{1 \to 2,\ 2 \to 3,\ 3 \to 1\}$
$\mathcal{S}_{L_2} = (\overline{K}_{L_2}, D_{L_2})$ $\quad \overline{K}_{L_2}$ formed by cells $\{\gamma_1, \gamma_2, \gamma_4\}$ and t $D_{L_2} = \{1 \to 2,\ 2 \to 4,\ 4 \to 8\}$
$\mathcal{S}_{L_3} = (\overline{K}_{L_3}, D_{L_3})$ $\quad \overline{K}_{L_3}$ formed by cells $\{\gamma_6, \gamma_7, \gamma_8\}$ and $D_{L_3} = \{8 \to 7,\ 7 \to 6,\ 6 \to 8\}$
$\mathcal{S}_{L_4} = (\overline{K}_{L_4}, D_{L_4})$ $\quad \overline{K}_{L_4}$ formed by cells $\{\gamma_5, \gamma_7, \gamma_8\}$ and $D_{L_4} = \{8 \to 7,\ 7 \to 5,\ 5 \to 1\}$

The two weak stripexes, namely, $\mathcal{S}_{L_2}$ and $\mathcal{S}_{L_4}$, have local twists, while the strong stripexes $\mathcal{S}_{L_1}$ and $\mathcal{S}_{L_3}$ do not (Fig. 9.9). If the subcomplex associated to the degenerated generatex ($\mathcal{G}_{L_2}$) is analyzed as an independent complex, there are no torsion groups.

9.5 Application to a Delay Differential Equations System

Biological and natural systems often rely on feedback processes that are typically formulated as Delay Differential Equations (DDEs). A priori, such systems have an infinite-dimensional phase space, as opposed to ordinary differential equations (ODEs) that are finite dimensional by definition.

There is a myriad of examples of DDEs models in the literature, some of which are proposed for understanding how tumor growth can be influenced by delayed interactions between cancer cells and the microenvironment [45]. This is studied with models that qualitatively describe the interactions between host, immune, and tumor cells at the tissue level in correspondence with various clinical features [45].

Clinical evidences showing that some of these interactions can be delayed in time [46, 47] encouraged the study of a set of three differential equations with two time delays. The first delay τ_1 describes how tumor cells inhibit the activity of immune cells, and the second delay τ_2 corresponds to the inhibition of tumor cells by effector cells. This can be written as follows:

$$\begin{aligned}
\dot{E} &= \frac{\rho T E}{g+T} - \beta_1 T\left(t-\tau_1\right) E\left(t-\tau_1\right) - \delta E \\
\dot{H} &= \alpha H\left(1-\frac{H}{k_1}\right) - \gamma_1 T H \\
\dot{T} &= a T\left(1-\frac{T}{k_2}\right) - \beta_2 T\left(t-\tau_2\right) E\left(t-\tau_2\right) - \gamma_2 T H,
\end{aligned}$$

where $E(t)$, $H(t)$, and $T(t)$ designate the population of activated effector cells, host cells, and tumor cells, respectively. Normalization of the state variables using

$$(x, y, z) = \left(\frac{E}{g}, \frac{H}{k_1}, \frac{T}{k_2}\right)$$

as well as the involved parameters produces the normalized dynamical system that we will be considering in this section.

$$\begin{aligned}
\dot{x} &= \frac{\rho x z}{g+z} - \beta_1 x\left(t-\tau_1\right) z\left(t-\tau_1\right) - \delta x \\
\dot{y} &= \alpha y(1-y) - \gamma_1 y z \\
\dot{z} &= z(1-z) - \beta_2 x\left(t-\tau_2\right) z\left(t-\tau_2\right) - \gamma_2 y z.
\end{aligned}$$

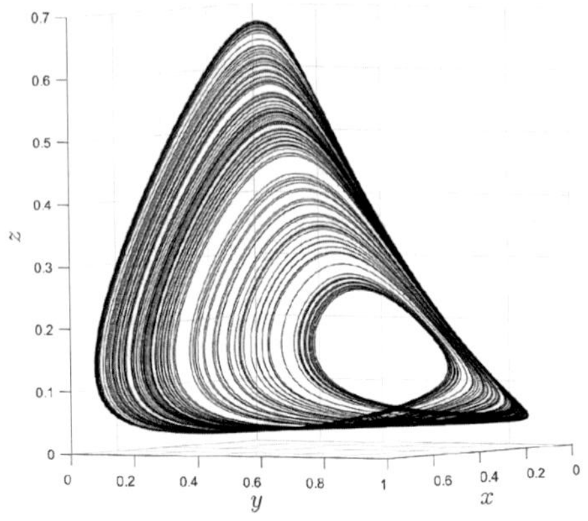

Fig. 9.10 The attractor generated by the DDEs system for the state variables x, y, z and parameter values $\rho = 4.5$, $\delta = 0.5$, $\alpha = 0.5$, $g = 1.0$, $\gamma_1 = 1.5$, $\gamma_2 = 1.0$, $\beta_1 = 0.2$, $\beta_1 = 2.5$). The time delays are $\tau_1 = 0.012$ and $\tau_2 = 0.012$

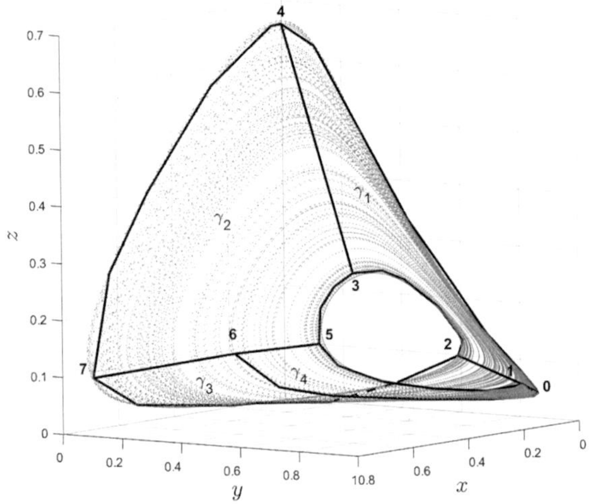

Fig. 9.11 Simplified BRAMAH cell complex K_{DDE} for the attractor in Fig. 9.10

The chaotic attractor produced by the normalized DDE system is shown in Fig. 9.10. The integration is done with the following initial conditions $x(0) = 0.1$, $y(0) = 0.2$, $z(0) = 0.5$, and parameter values $\rho = 4.5$, $\tau_1 = 0.012$, $\tau_2 = 0.1$ $g = 1$, $\alpha = 0.5$, $\beta_1 = 0.2$, $\beta_2 = 2.5$, $\gamma_1 = 1.5$, $\gamma_2 = 1.0$, $\delta = 0.5$.

The template for this attractor is shown in Fig. 8 in [45]. It is found to have two strips, which can be described by the same linking matrix as the Rössler attractor. This result indicates that the delayed system is a priori equivalent to a set of ODEs exhibiting spiral type chaos without delays.

Let us now use the new elements for a theory of chaos topology developed in this chapter, starting with the point cloud obtained from the DDE system. A BRAMAH complex for the chaotic solution is shown in Fig. 9.11. This simple 2-complex labeled K_{DDE} with as few as four 2-cells can be "unfolded" in the form of a planar diagram, where repeated indices indicate the gluing instructions (Fig. 9.12 left).

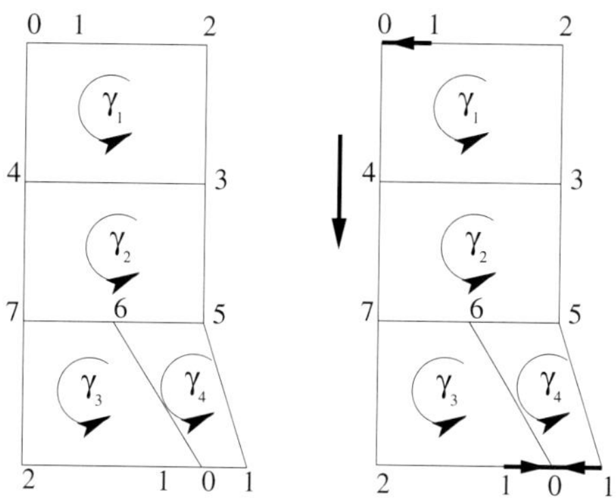

Fig. 9.12 Planar diagrams of the BRAMAH complex K_{DDE} and of the flow-oriented version $\overline{K}_{\text{DDE}}$. The thick lines mark the joining 1-cells that are now oriented in agreement with the flow-oriented 2-cells. The direction of the flow is indicated with the side arrow

DDE Branched Manifold

The 2-branched manifold in the 3-dimensional phase space supporting the chaotic solution of the DDE system can be described in terms of the properties of K_{DDE}. The DDE BRAMAH complex is homologically equivalent to the Rössler case: $H_0(K_{\text{DDE}}) \sim \mathbb{Z}^1$, $H_1(K_{\text{DDE}}) \sim \mathbb{Z}^1$, and $H_2(K_{\text{DDE}}) \sim \emptyset$. The generator of the first homology group is given the following chain: $+\langle 1, 2\rangle - \langle 1, 5\rangle + \langle 2, 3\rangle + \langle 3, 5\rangle$. The Rössler and DDE systems are homologically equivalent.

There is however an important difference between them that can be detected when computing the border of the sum of all the 2-cells of the complex K_{DDE}. The Rössler complex is torsion-free, while K_{DDE} is not, and one can in fact see that $\langle 1, 2\rangle$ appears with a factor -2, indicating that the basic strip structure into which reinjection will occur is not a standard band (or cylinder) as in the Rössler attractor, but a Möbius band.

Concerning the branching of the complex, $\langle 0, 1\rangle$ is a joining 1-cell, and γ_1, γ_3, and γ_4 are joining 2-cells.

DDE Stripexes

The flow-oriented complex ($\overline{K}_{\text{DDE}}$) is shown in Fig. 9.12 (right). The digraph related to $\overline{K}_{\text{DDE}}$ is given by $\{1 \to 2,\ 2 \to 3,\ 2 \to 4,\ 3 \to 1,\ 4 \to 1\}$. It is not shown in a new figure because it is graphically identical to D_R in Fig. 9.6.

The templex $T_{\text{DDE}} = (\overline{K}_{\text{DDE}}, D_{\text{DDE}})$ for the delay differential equation system has two simple generatexes $\{\mathcal{G}_{\text{DDE}_1}, \mathcal{G}_{\text{DDE}_2}\}$. There are therefore two strong stripexes $\{\mathcal{S}_{\text{DDE}_1}, \mathcal{S}_{\text{DDE}_2}\}$.

$$\begin{aligned}
\mathcal{G}_{\text{DDE}_1} = \mathcal{S}_{\text{DDE}_1} &= (\overline{K}_{\text{DDE}_1}, D_{\text{DDE}_1}) \quad \overline{K}_{\text{DDE}_1} \text{ formed by cells } \{\gamma_1,\ \gamma_2,\ \gamma_4\} \text{ and} \\
D_{\text{DDE}_1} &= \{1 \to 2,\ 2 \to 4,\ 4 \to 1\} \\
\mathcal{G}_{\text{DDE}_2} = \mathcal{S}_{\text{DDE}_2} &= (\overline{K}_{\text{DDE}_2}, D_{\text{DDE}_2}) \quad \overline{K}_{\text{DDE}_2} \text{ formed by cells } \{\gamma_1,\ \gamma_2,\ \gamma_3\} \text{ and} \\
D_{\text{DDE}_2} &= \{1 \to 2,\ 2 \to 3,\ 3 \to 1\}
\end{aligned}$$

Only the second stripex $\mathcal{S}_{\text{DDE}_2}$ has a twist (Fig. 9.13). If the subcomplexes forming the two stripexes are analyzed as independent complexes, they have the properties of a standard band (cylinder) and of a Möbius band, respectively.

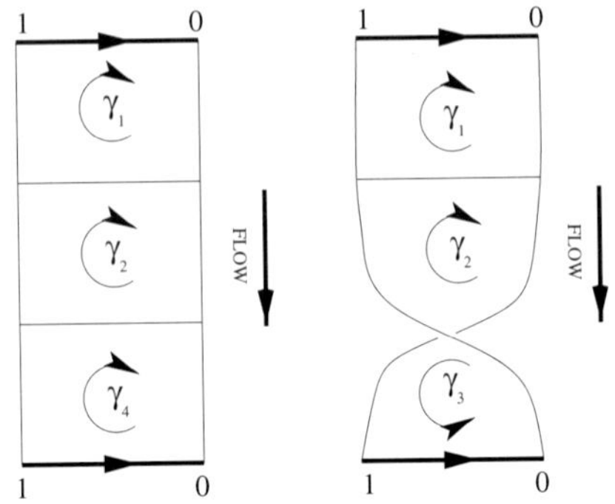

Fig. 9.13 Subcomplexes $\overline{K}_{DDE_1}$ (left), $\overline{K}_{DDE_2}$ (right) corresponding to stripexes $\mathcal{S}_{DDE_1}$ and $\mathcal{S}_{DDE_2}$, respectively

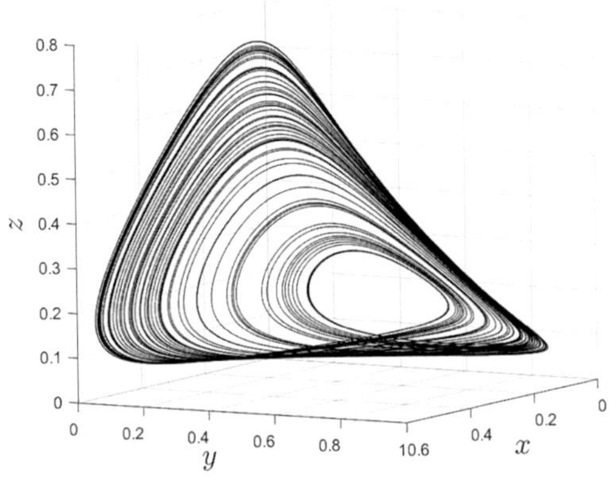

Fig. 9.14 The chaotic attractor generated with the ODE whose templex is equivalent to $\mathcal{T}_{\text{DDE}}$. Initial conditions are set to $x(0) = 0.1$, $y(0) = 0.1$, $z(0.01)$

Looking for an ODE system whose topological properties reproduce those of the DDE system is possible using global modeling techniques [45, 48]. This search yields:

$$\dot{x} = -0.407x + 2.780\frac{xz}{1+z}$$
$$\dot{y} = 0.533(1 - 0.9456y) - 1.497yz$$
$$\dot{z} = 0.753(1 - 0.8486z) - 0.703tz - 2.323xz.$$

If we integrate this ODE system and we analyze it using a templex, we obtain, once again, a complex and a digraph whose joint properties match those of the DDE system.

It is also possible to conclude that the dynamics of the DDE model and of the equivalent ODE system have some but not all the topological properties of the Rössler spiral chaos. The delayed interactions between cancer cells and the microenvironement can be modeled without delays, but not without the orientability properties of the branched manifold upon which the two stripexes are inscribed (Fig. 9.14).

9.6 Hopes for the Future

A decade ago, Robert Gilmore—in the first chapter of the book celebrating his 70th birthday [49]—wrote a paragraph with Christophe Letllier, having the same title we now give to this final section of our chapter. There they asked:

> Is it possible to find useful topological structures to replace the knots that have been fundamental for the development of the topological analysis program for chaotic attractors in $\mathbb{R}^3$?

They imagined that the solution might come from skipping the first, knotty step, even the second, template step, to jump right to the end of the chain and find some way to make statements about the mechanisms that create chaotic behavior.

The new elements presented in this chapter show that (i) approximating the dataset in phase space with properly glued cells that form a topologically faithful skeleton of the underlying branched manifold, and (ii) describing the flow across such cells with a directed graph, provides an affirmative answer to the question posed in 2013.

A templex describes the manner in which a flow is structured in phase space using the strength of algebraic topology and graph theory to incorporate the information of the nonequivalent paths along a nontrivial topological structure. The theory is not limited to chaotic attractors: These definitions can be applied to different types of flows in phase space. It can also describe quasi-periodic solutions, providing an intelligible difference between them in terms of the mathematical concepts that are proposed to describe them.

Returning to the wishes for the future of the 2013 book that have not yet been accomplished, it would be extraordinary to have theorems for templexes and flows that play the role of the theorems for mappings that form the basis of catastrophe theory [50]. The object formed by the cell complex and the directed graph is newly born and invites topologists to dive into its properties. The examples shown here are all three-dimensional systems, but four-dimensional cases are discussed in [9, 12]. Scaling upward in the phase space dimension of the dynamical systems is an active purpose within the templex-based topological program.

The content of this chapter has been devoted to deterministic systems governed by ODEs or DDEs, but the interested reader may also wish to discover how a templex can be extended to a stochastic framework [8, 51, 52] and used to describe a noise-driven dynamical system whose topological structure in phase space is no longer an invariant.

Acknowledgments The content of this chapter relies to a great extent on recent joint work with Christophe Letellier and Michael Ghil [12, 41, 51, 52]. It is a pleasure to thank them for their respective inputs into the work published in these cited articles. We also thank Iván Sadofschi Costa and Sylvain Mangiarotti for stimulating discussions. This work was supported by the French National program LEFE (Les Enveloppes Fluides et l'Environnement) and by the CLIMAT-AMSUD 21-CLIMAT-05 project (D.S.). G.D.C. is grateful for her postdoctoral scholarship from CONICET. TeMPlex ANR Project AAP CE56, PI Denisse Sciamarella, is also acknowledged.

References

1. E. Lorenz, Deterministic nonperiodic flow. J. Atmos. Sci. **20**, 130–141 (1963)
2. W. Tucker, A rigorous ODE solver and Smale's 14th problem. Found. Comput. Math. **2**, 53–117 (2002)
3. I. Ovsyannikov, D. Turaev, Analytic proof of the existence of the Lorenz attractor in the extended Lorenz model. Nonlinearity **30**, 115 (2016)
4. C. Letellier, O. Rössler, *Chaos: The World of Nonperiodic Oscillations* (Springer, Berlin, 2020)
5. A. Groth, M. Ghil, Multivariate singular spectrum analysis and the road to phase synchronization. Phys. Rev. E **84**, 036206 (2011)
6. A. Gierzkiewicz, P. Zgliczyński, Periodic orbits in the Rössler system. Commun. Nonlinear Sci. Numer. Simul. **101**, 105891 (2021)
7. O. Rössler. The chaotic hierarchy. Zeitschrift Für Naturforschung A **38**, 788–801 (1983)
8. M. Ghil, D. Sciamarella, Dynamical systems, algebraic topology, and the climate sciences. EGUsphere **2023**, 1–53 (2023)
9. D. Sciamarella, G. Mindlin, Unveiling the topological structure of chaotic flows from data. Phys. Rev. E **64**, 036209 (2001)
10. G. Charó, G. Artana, D. Sciamarella, Topology of dynamical reconstructions from Lagrangian data. Physica D **405**, 132371 (2020)
11. G. Charó, G. Artana, D. Sciamarella, Topological colouring of fluid particles unravels finite-time coherent sets. J. Fluid Mech. **923**, A17 (2021)
12. G. Charó, C. Letellier, D. Sciamarella, Templex: a bridge between homologies and templates for chaotic attractors. Chaos: An Interdisciplinary J. Nonlinear Sci. **32**, 083108 (2022)
13. C. Letellier, V. Messager, Influences on Otto E. Rössler's earliest paper on chaos. Int. J. Bifurcation Chaos **20**, 3585–3616 (2010)
14. T. Palmer, A. Döring, G. Seregin, The real butterfly effect. Nonlinearity **27**, R123 (2014)
15. H. Poincaré, *Les méthodes nouvelles de la mécanique céleste*, vol. 3 (Gauthier-Villars, France, 1992)
16. D. Auerbach, J. éCvitanovi, G.G. Eckmann, I. Procaccia, Exploring chaotic motion through periodic orbits. Phys. Rev. Lett. **58**, 2387–2389 (1987)
17. C. Grebogi, E. Ott, J. Yorke, Unstable periodic orbits and the dimensions of multifractal chaotic attractors. Phys. Rev. A **37**, 1711–1724 (1988)
18. D. Lathrop, E. Kostelich, Characterization of an experimental strange attractor by periodic orbits. Phys. Rev. A **40**, 4028 (1989)
19. G. Mindlin, R. Gilmore, Topological analysis and synthesis of chaotic time series. Physica D **58**, 229–242 (1992)
20. R. Williams, The structure of Lorenz attractors. Publ. Math. De L'I.H.E.S. **50**, 73–99 (1979)
21. J. Birman, R. Williams, Knotted periodic orbits in dynamical systems. Lorenz's equations. Topology **22**, 47–82 (1983)
22. M. Muldoon, R. MacKay, J. Huke, D. Broomhead, Topology from time series I. Physica D **65**, 1–16 (1993)
23. É. Ghys, Knots and dynamics. Int. Congr. Math. **1**, 247–277 (2007)
24. J. Birman, The mathematics of Lorenz knots, in *Topology and Dynamics of Chaos: In Celebration of Robert Gilmore's 70th Birthday* (2013), pp. 127–148
25. J. Guckenheimer, R. Williams, Structural stability of Lorenz attractors. Publ. Math. De L'IHÉS **50**, 59–72 (1979)
26. R. Gilmore, How topology came to chaos, in *Topology and Dynamics of Chaos: In Celebration of Robert Gilmore's 70th Birthday* (2013), pp. 127–148
27. P. Melvin, N. Tufillaro, Templates and framed braids. Phys. Rev. A **44**, R3419 (1991)
28. N. Tufillaro, T. Abbott, J. Reilly, *An Experimental Approach to Nonlinear Dynamics and Chaos* (Addison-Wesley, New York, 1992)
29. C. Letellier, P. Dutertre, B. Maheu, Unstable periodic orbits and templates of the Rössler system: toward a systematic topological characterization. Chaos **5**, 271–282 (1995)

30. C. Letellier, N. Stankevich, O. Rössler, Dynamical taxonomy: some taxonomic ranks to systematically classify every chaotic attractor. Int. J. Bifurcation Chaos **32**, 2230004 (2022)
31. G. Boulant, M. Lefranc, S. Bielawski, D. Derozier, A nonhorseshoe template in a chaotic laser model. Int. J. Bifurcation Chaos **8**, 965–975 (1998)
32. C. Letellier, M. Aziz-Alaoui, Analysis of the dynamics of a realistic ecological model. Chaos, Solitons Fractals **13**, 95–107 (2002)
33. C. Letellier, L. Aguirre, J. Maquet, M. Aziz-Alaoui, Should all the species of a food chain be counted to investigate the global dynamics?. Chaos, Solitons Fractals **13**, 1099–1113 (2002)
34. L. Aguirre, C. Letellier, J. Maquet, Forecasting the time series of sunspot numbers. Sol. Phys. **249**, 103–120 (2008)
35. J. Used, J. Martín, Reverse horseshoe and spiral templates in an erbium-doped fiber laser. Phys. Rev. E **79**, 046213 (2009)
36. C. Letellier, F. Denis, L. Aguirre, What can be learned from a chaotic cancer model?. J. Theor. Biol. **322**, 7–16 (2013)
37. C. Letellier, S. Mangiarotti, L. Minati, M. Frasca, J. Barbot, Optimal placement of sensor and actuator for controlling low-dimensional chaotic systems based on global modeling. Chaos: An Interdisciplinary J. Nonlinear Sci. **33**, 013140 (2023)
38. M. Natiello, et al., *The User's Approach to Topological Methods in 3d Dynamical Systems* (World Scientific, Singapore, 2007)
39. L. Kinsey, Topology of surfaces (Springer, Berlin, 1993)
40. D. Sciamarella, G. Mindlin, Topological structure of chaotic flows from human speech data. Phys. Rev. Lett. **64**, 1450–1453 (1999)
41. G. Charó, D. Sciamarella, S. Mangiarotti, G. Artana, C. Letellier, Observability of laminar bidimensional fluid flows seen as autonomous chaotic systems. Chaos: An Interdisciplinary J. Nonlinear Sci. **29**, 123–126 (2019)
42. J. Guckenheimer, P. Holmes, *Nonlinear Oscillations, Dynamical Systems, and Bifurcations of Vector Fields* (Springer, Berlin, 1983)
43. O. Ménard, C. Letellier, J. Maquet, L. Sceller, G. Gouesbet, Analysis of a non synchronized sinusoidally driven dynamical system. Int. J. Bifurcation Chaos **10**, 1759–1772 (2000)
44. M. Rosalie, C. Letellier, Systematic template extraction from chaotic attractors: I. Genus-one attractors with an inversion symmetry. J. Phys. A **46**, 375101 (2013)
45. D. Ghosh, S. Khajanchi, S. Mangiarotti, F. Denis, S. Dana, C. Letellier, How tumor growth can be influenced by delayed interactions between cancer cells and the microenvironment?. BioSystems **158**, 17–30 (2017)
46. J. Brahmer, S. Tykodi, L. Chow, W. Hwu, S. Topalian, P. Hwu, C. Drake, L. Camacho, J. Kauh, K. Odunsi, et al., Safety and activity of anti–PD-L1 antibody in patients with advanced cancer. N. Engl. J. Med. **366**, 2455–2465 (2012)
47. J. Brahmer, K. Reckamp, P. Baas, L. Crinò, W. Eberhardt, E. Poddubskaya, S. Antonia, A. Pluzanski, E. Vokes, E. Holgado, et al., Nivolumab versus docetaxel in advanced squamous-cell non–small-cell lung cancer. N. Engl. J. Med. **373**, 123–135 (2015)
48. L. Aguirre, C. Letellier, Modeling nonlinear dynamics and chaos: a review. Math. Probl. Eng. **2009**, 238960 (2009)
49. C. Letellier, R. Gilmore, Introduction to topological analysis. Topology and Dyn. Chaos Celebration of Robert Gilmore's 70th Birthday **84**, 1–19 (2013)
50. V. Arnol'd, *Catastrophe Theory* (Springer, Berlin, 1986)
51. G. Charó, M. Chekroun, D. Sciamarella, M. Ghil, Noise-driven topological changes in chaotic dynamics. Chaos **31**, 103115 (2021)
52. G. D. Charó, M. Ghil, D. Sciamarella, Random templex encodes topological tipping points in noise-driven chaotic dynamics. Chaos: An Interdisciplinary Journal of Nonlinear Science, **33**(10) (2023).

SPRINGER NATURE

GPSR Compliance

The European Union's (EU) General Product Safety Regulation (GPSR) is a set of rules that requires consumer products to be safe and our obligations to ensure this.

If you have any concerns about our products, you can contact us on ProductSafety@springernature.com

In case Publisher is established outside the EU, the EU authorized representative is:

Springer Nature Customer Service Center GmbH
Europaplatz 3
69115 Heidelberg, Germany

The manufacturer's authorised representative in the EU is Springer Nature Customer Service Centre GmbH, Europaplatz 3, 69115 Heidelberg, Germany. If you have any concerns regarding our products, please contact ProductSafety@springernature.com

Printed and bound by CPI Group (UK) Ltd, Croydon, CR0 4YY
13/10/2026
02232444-0001